T0309511

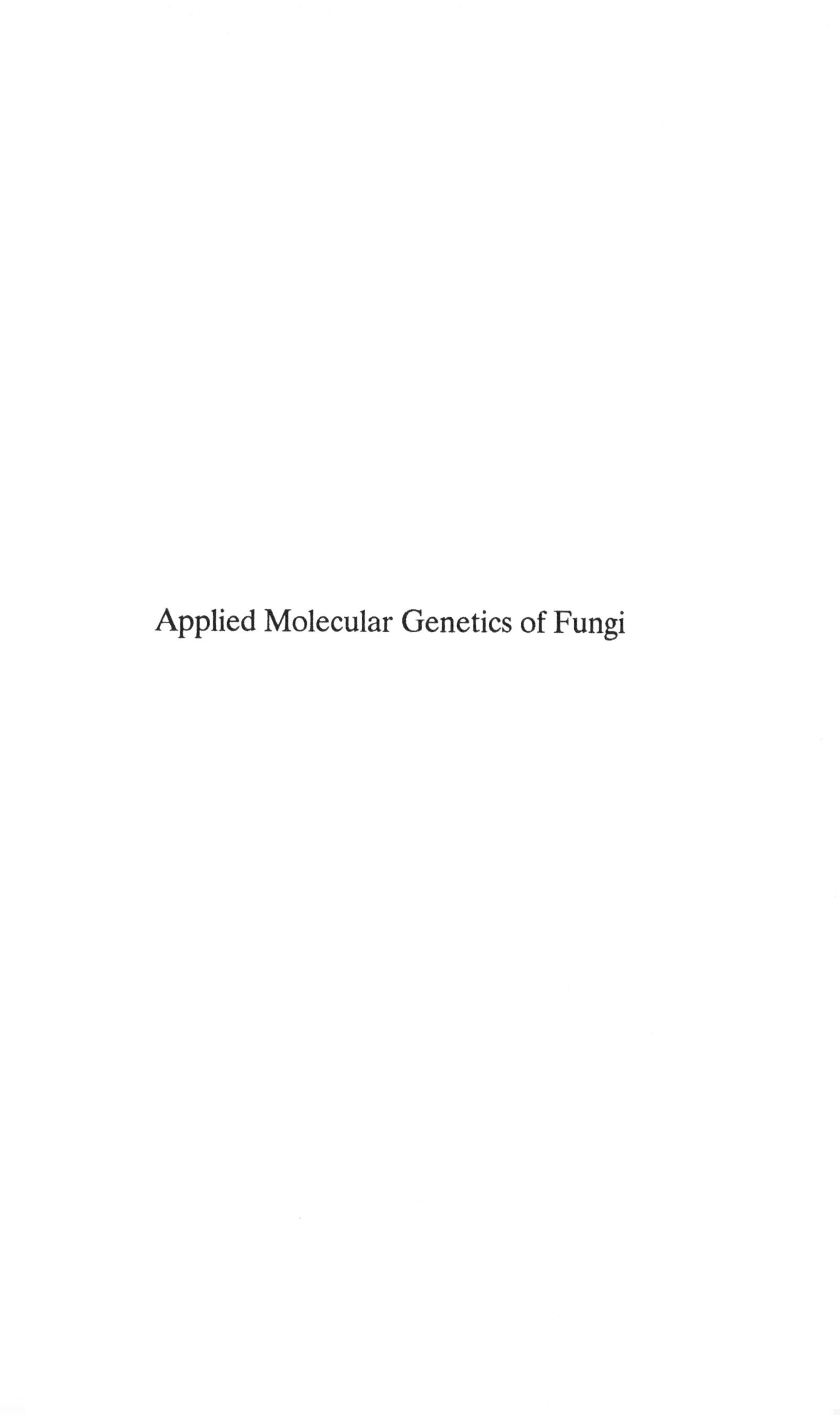

Applied Molecular Genetics of Fungi

Applied Molecular Genetics of Fungi

Symposium of the British Mycological Society held at the University of Nottingham, April 1990

Edited by
J. F. Peberdy, C. E. Caten, J. E. Ogden & J. W. Bennett

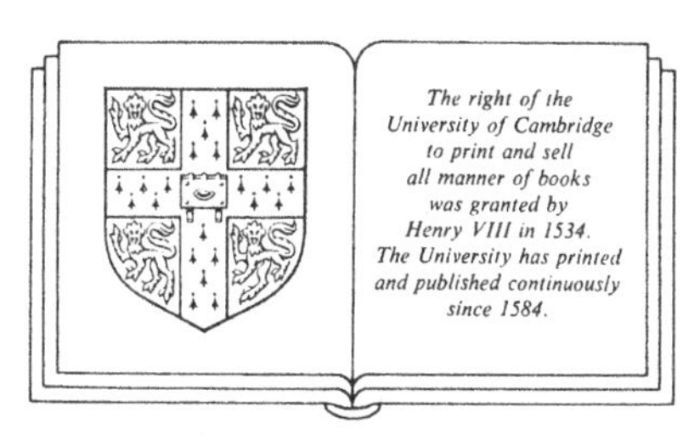

Published for the British Mycological Society by
CAMBRIDGE UNIVERSITY PRESS
Cambridge
New York Port Chester Melbourne Sydney

CAMBRIDGE UNIVERSITY PRESS
Cambridge, New York, Melbourne, Madrid, Cape Town, Singapore, São Paulo

Cambridge University Press
The Edinburgh Building, Cambridge CB2 8RU, UK

Published in the United States of America by Cambridge University Press, New York

www.cambridge.org
Information on this title: www.cambridge.org/9780521415712

© British Mycological Society 1991

First published 1991

Designed and produced by **Page Design,** 48 Alan Road, Stockport SK4 4LE

A catalogue record for this publication is available from the British Library

ISBN 978-0-521-41571-2 hardback

Transferred to digital printing 2007

Contents

Contributors *page vi*

Preface *ix*

1 Gene-transfer systems and vector development for filamentous fungi 1
C. A. M. J. J. van den Hondel & P. J. Punt

2 Strategies for cloning genes from filamentous fungi 29
G. Turner

3 Novel methods of DNA transfer 44
J. W. Watts & N. J. Stacey

4 *Saccharomyces cerevisiae*: a host for the production of foreign proteins 66
J. E. Ogden

5 The molecular biology of *Trichoderma reesei* and its application to biotechnology 85
M. Penttilä, T. T. Teeri, H. Nevalainen & J. K.C. Knowles

6 Expression of heterologous genes in filamentous fungi 103
R. W. Davies

7 Methylotrophic yeasts as gene expression systems 118
R. A. Veale & P. E. Sudbery

8 Strain improvement of brewing yeast 128
E. Hinchliffe

9 Identification of the *Cephalosporium acremonium pbc*AB gene using predictions from an evolutionary hypothesis 145
P. L. Skatrud, J. Hoskins, M. B. Tobin, J. R. Miller, J. S. Wood, S. Kovacevic & S. W. Queener

10 Applications of genetically manipulated yeasts 160
A. W. M. Strasser, Z. A. Janowicz, R. O. Roggenkamp, U. Dahlems, U. Weydemann, A. Merckelbach, G. Gellissen, R. J. Dohmen, M. Piontek, K. Melber & C. P. Hollenberg

11 Molecular biology of fungal plant pathogenicity 170
R. P. Oliver, M. L. Farman, N. J. Talbot & M. T. McHale

Index 183

Contributors

Joan W. Bennett *Department of Cell & Molecular Biology, 2000 Percival Stern Hall, Tulane University, New Orleans, Louisiana 70118, U.S.A.*

Christopher E. Caten *School of Biological Sciences, University of Birmingham, PO Box 363, Birmingham B15 2TT, U.K.*

Ulrike Dahlems *Rhein Biotech GmbH, Erkratherstrasse 230, 4000 Düsseldorf 1, Germany*

R. Jürgen Dohmen *Institut für Mikrobiologie, Heinrich-Heine-Universität Düsseldorf, Universitätstrasse 1, 4000 Düsseldorf 1, Germany*

R. Wayne Davies *Robertson Institute of Biotechnology, Department of Genetics, University of Glasgow, Glasgow G11 5JS, U.K.*

Mark L. Farman *Norwich Molecular Plant Pathology Group, School of Biological Sciences, University of East Anglia, Norwich NR4 7TJ, U.K.*

Gerd Gellissen *Rhein Biotech GmbH, Erkratherstrasse 230, 4000 Düsseldorf 1, Germany*

Edward Hinchliffe *Bass Brewers Ltd, 137 High Street, Burton-on-Trent, E. Staffordshire DE14 1JZ, U.K.*

Cornelius P. Hollenberg *Institut für Mikrobiologie, Heinrich-Heine-Universität Düsseldorf, Universitätstrasse 1, 4000 Düsseldorf 1, Germany*

JoAnn Hoskins *Molecular Genetics Department, Lilly Research Laboratories, Lilly Corp. Center, Indianapolis, Indiana 46285, U.S.A.*

Zbigniew A. Janowicz *Rhein Biotech GmbH, Erkratherstrasse 230, 4000 Düsseldorf 1, Germany*

Jonathan K. C. Knowles *VTT Biotechnical Laboratory, POB 202, SF-02151 Espoo, Finland*

Steven Kovacevic *Molecular Genetics Department, Lilly Research Laboratories, Lilly Corp. Center, Indianapolis, Indiana 46285, U.S.A.*

Mark T. McHale *Norwich Molecular Plant Pathology Group, School of Biological Sciences, University of East Anglia, Norwich NR4 7TJ, U.K.*

Karl Melber *Rhein Biotech GmbH, Erkratherstrasse 230, 4000 Düsseldorf 1, Germany*

Armin Merckelbach *Rhein Biotech GmbH, Erkratherstrasse 230, 4000 Düsseldorf 1, Germany*

James R. Miller *Molecular Genetics Department, Lilly Research Laboratories, Lilly Corp. Center, Indianapolis, Indiana 46285, U.S.A.*

Helena Nevalainen *Research Laboratories, Alko Ltd., POB 350, SF-00101 Helsinki, Finland*

Jill E. Ogden *Delta Biotechnology Ltd., Castle Court, Castle Boulevard, Nottingham NG7 1FD, U.K.*

Richard P. Oliver *Norwich Molecular Plant Pathology Group, School of Biological Sciences, University of East Anglia, Norwich NR4 7TJ, U.K.*

John F. Peberdy *Department of Botany, School of Biological Sciences, University of Nottingham, University Park, Nottingham NG7 2RD, U.K.*

Merja Penttilä *VTT Biotechnical Laboratory, POB 202, SF-02151 Espoo, Finland*

Michael Piontek *Institut für Mikrobiologie, Heinrich-Heine-Universität Düsseldorf, Universitätstrasse 1, 4000 Düsseldorf 1, Germany*

Peter J. Punt *TNO Medical Biological Laboratory, PO Box 45, 2280 AA Rijswijk, The Netherlands*

Stephen W. Queener *Molecular Genetics Department, Lilly Research Laboratories, Lilly Corp. Center, Indianapolis, Indiana 46285, U.S.A.*

Rainer O. Roggenkamp *Institut für Mikrobiologie, Heinrich-Heine-Universität Düsseldorf, Universitätstrasse 1, 4000 Düsseldorf 1, Germany*

Paul L. Skatrud *Molecular Genetics Department, Lilly Research Laboratories, Lilly Corp. Center, Indianapolis, Indiana 46285, U.S.A.*

Nicholas J. Stacey *John Innes Centre for Plant Science Research, Colney Lane, Norwich NR4 7UH, U.K.*

Alexander W. M. Strasser *Rhein Biotech GmbH, Erkratherstrasse 230, 4000 Düsseldorf 1, Germany*

Peter E. Sudbery *Department of Molecular Biology and Biotechnology, University of Sheffield, Sheffield S10 2TN, U.K.*

Nicholas J. Talbot *Norwich Molecular Plant Pathology Group, School of Biological Sciences, University of East Anglia, Norwich NR4 7TJ, U.K.*

Tuula T. Teeri *VTT Biotechnical Laboratory, POB 202, SF-02151 Espoo, Finland*

Matthew B. Tobin *Molecular Genetics Department, Lilly Research Laboratories, Lilly Corp. Center, Indianapolis, Indiana 46285, U.S.A.*

Geoffrey Turner *Department of Molecular Biology and Biotechnology, The University, Sheffield S10 2TN, U.K.*

Cees A. M. J. J. van den Hondel *TNO Medical Biological Laboratory, PO Box 45, 2280 AA Rijswijk, The Netherlands*

Rosemary A. Veale *ICI Joint Laboratories, Department of Biochemistry, University of Leicester, Leicester LE1 7RH, U.K.*

John W. Watts *John Innes Centre for Plant Science Research, Colney Lane, Norwich NR4 7UH, U.K.*

Ulrike Weydemann *Rhein Biotech GmbH, Erkratherstrasse 230, 4000 Düsseldorf 1, Germany*

John S. Wood *Molecular Genetics Department, Lilly Research Laboratories, Lilly Corp. Center, Indianapolis, Indiana 46285, U.S.A.*

Preface

The interactions of fungi with mankind are both beneficial and harmful and are deeply rooted in the history of human society and agriculture. Over the centuries man has sought to manipulate the growth of fungi to his advantage; the methods used though largely empirical have often been highly successful. Since the initial development of recombinant DNA technology in bacteria in the early 1970s, biology has been undergoing a revolution which is spreading to all organisms, including fungi. This revolution is marked by the emergence of a new discipline, molecular biology, at the interface between biochemistry and genetics. The approach and techniques of molecular biology enable us to ask and answer fundamental questions about many aspects of fungal biology, and open the way to the directed manipulation of fungal metabolism.

This book arises from a symposium on 'Fungal Molecular Biology' held by the British Mycological Society at the University of Nottingham in April 1990. Altogether, there were 29 main papers presented at the symposium, covering a broad range of both fundamental and applied aspects of fungal molecular biology. In considering a book based on the meeting it seemed desirable, given the inevitable restrictions on space and cost, to focus on one or two areas. The editors decided to highlight the rapid development of gene transfer and cloning techniques in fungi and the ways in which these are being exploited in species of economic importance either in biotechnology or as plant pathogens. The 11 contributions in this volume were selected on that basis.

The relevant methodologies for gene manipulations in fungi are described in the first three chapters. In chapter 1 (Van den Hondel & Punt) the development of suitable vectors and gene transfer systems for filamentous fungi discussed and the wide applicability of these techniques to all fungi is clearly established. One point that emerges is that although a basis of classical genetics is useful, it is not essential. A central feature of this new approach to genetic manipulation is the cloning of genes; several strategies are available in filamentous fungi and the most applicable in each situation can be readily identified (Chapter 2, Turner). To date, the technology for introducing vectors into fungal cells has been restricted primarily to systems based on polyethylene glycol promotion of DNA uptake into protoplasts. Workers manipulating plant and animal cells have explored more 'dramatic' procedures as described by Watts & Stacey (Chapter 3).

Not surprisingly, progress in yeast molecular biology has been even more rapid than that with filamentous fungi. Several contributions concerning yeast research were included in the symposium to provide a point of reference for possible future developments with the filamentous fungi. Advances with *Saccharomyces cerevisiae* stem, in part, from its importance in brewing, where several opportunities for exploitation of recombinant strains exist (Chapter 8, Hinchliffe), but mainly from previously established fundamental knowledge of biochemistry, cell biology and genetics in this organism. A clear example of building on the latter is the use of *Saccharomyces* as a host for the expression of heterologous proteins (Chapter 4, Ogden). Despite the fact that this fungus secretes only a limited range of proteins naturally, it can be engineered genetically to secrete significant amounts of recombinant proteins. The success with *Saccharomyces* prompted interest in several other yeasts including the methylotrophic species and several systems are now operational (Chapter 7, Veale & Sudbery; Chapter 10, Strasser *et al.*).

Industrially, the filamentous fungi are best known as sources of antibiotics, organic acids and enzymes. Several of the genes encoding biosynthetic enzymes for β-lactam synthesis have been cloned and manipulated; the advances made in this area in *Cephalosporium* (*Acremonium*) are considered by Skatrud *et al.* (Chapter 9). *Trichoderma* species are used commercially as the producers of a range of hydrolytic enzymes which are secreted into the growth medium. The cellulase system has been investigated using molecular genetic techniques and this has led not only to improvements in cellulase production, but also to the exploitation of this fungus as a host for the expression of heterologous proteins (Chapter 5, Penttilä *et al.*). The Aspergilli are of particular importance in fungal molecular biology as they contain both model experimental and industrially important species. Several of these species are the subject of intense study aimed at developing them as hosts for the commercial production of mammalian proteins (Chapter 6, Davies).

The detrimental economic effects of fungi as agents of plant disease are of even greater importance than the beneficial role of fungi in biotechnology. Most phytopathogenic fungi are not amenable to study by the classical methods of genetics and biochemistry and, as a result, the basic mechanism of fungal pathogen-plant host interactions are poorly understood. However, the approach and techniques of molecular genetics bypass many of these difficulties and are transforming knowledge of all aspects of the biology of these fungi. Clearly there is along way to go before we understand the molecular basis of fungal pathogenicity, but sound foundations are being laid as described in the final chapter (Chapter 11, Oliver *et al.*).

The editors of this volume are grateful to the British Mycological Society for providing the means to organise such a timely and interesting symposium and for supporting the publication of this volume. Generous donations towards the costs of the symposium from Bicon Biochemicals, Cambridge University Press, Glaxo Group Research, Pfizer, SmithKline Beecham and Xenova are gratefully acknowledged. We wish to thank all those who contributed to the meeting and, in particular, the authors of the chapters in this volume for their cooperation in preparing the manuscripts for this book in as short a time as possible. Finally, special thanks go to David Moore and Page Design for their help, guidance and great efficiency in producing the book.

J. F. Peberdy
University of Nottingham

C. E. Caten
University of Birmingham

J. E. Ogden
Delta Biotechnology, Nottingham

J. W. Bennett
Tulane University, New Orleans

Chapter 1

Gene transfer systems and vector development for filamentous fungi

Cees A. M. J. J. van den Hondel & Peter J. Punt

Filamentous fungi have a number of properties which make them important both scientifically and economically. The economic importance can be illustrated by the large variety of products that are made by filamentous fungi, such as organic acids (e.g. citric acid), antibiotics (e.g. penicillin and cephalosporin) and numerous industrial enzymes (e.g. glucoamylase). Filamentous fungi are also used as food (mushrooms), food additives (e.g. the meat extender 'Quorn') and condiments (e.g. soy sauce). A severe, negative economic influence of filamentous fungi is their detrimental effect on crop yield. Plant pathogenic fungi cause annual crop losses of billions of pounds. In addition to their economic importance, filamentous fungi have interesting biological properties such as a complex life cycle, cell differentiation, highly regulated metabolic pathways and efficient secretion of proteins which make them attractive as a model for basic biological research of eukaryotic organisms.

In the pre-recombinant DNA period, physiological, biochemical and genetic studies were mainly carried out with *Neurospora crassa* and *Aspergillus nidulans*. Their haploid genomes, rapid life cycles, simple nutrient requirements and well developed genetic systems made them attractive model systems. Hence, it stands to reason that after the introduction of recombinant DNA techniques, systems for molecular genetic analysis were first developed in these intensively studied filamentous fungi. Thereafter, similar molecular techniques have been extended to less amenable species.

A prerequisite for molecular genetic research in filamentous fungi is the availability of a gene transfer system comprising a vector containing a selectable marker and a transformation procedure for introduction of the vector into the fungus. The specific properties of different types of selection markers can be used to design vectors for specific genetic manipulation strategies necessary for molecular genetic studies.

Recently, several excellent reviews have been published about transformation and genetic engineering of filamentous fungi (Fincham, 1989; Timberlake & Marshall, 1989; Goosen, Bos & Van den Broek, 1990). Therefore, in this chapter we will briefly summarise some of the features

Table 1.1. Overview of transformation systems used for filamentous fungi

Mycelial treatment	References
Protoplasts	
$CaCl_2$/PEG	Peberdy (1989) and references therein
liposomes	Radford *et al.* (1981)
electroporation	Ward *et al.* (1989); Thomas & Kenerly (1989); Goldman, Van Montagu & Herrera-Estrella (1990)
Intact cells	
Li acetate	Fincham (1989) and references therein; Bej & Perlin (1989)
biolistic	Armaleo *et al.* (1990)

of the gene transfer systems developed. Special attention will be given to some applications of these systems for genetic manipulation in *Aspergillus*.

Gene transfer systems

For genetic manipulation of filamentous fungi a gene transfer system is required that permits introduction of exogenous DNA and selection of those cells that have incorporated this DNA. This selection can be achieved by covalently linking the DNA to a vector which contains a selection marker. Both transformation frequency and type of transformant can be manipulated by using different types of vector.

Transformation procedure

The procedure to obtain DNA-mediated transformed fungal cells comprises the following steps:

- preparation of cells (protoplasts) which are competent to take up (vector) DNA
- treatment of these cells with the DNA
- regeneration of colony forming units
- selection/detection of those cells that have stably incorporated DNA.

A summary of the transformation systems used for filamentous fungi is given in Table 1.1.

Most frequently, protoplasts are used for the introduction of exogenous DNA. These protoplasts are obtained by incubation of mycelium or spores with cell wall-degrading enzymes in the presence of a compound that stabilizes the protoplasts (for an extensive overview of the different procedures and enzymes used, see Peberdy (1989)). Recently, transformation through electroporation of protoplasts was described (for references, see Table 1.1). Compared to the generally used $CaCl_2$/PEG method, no significant improvement of transformation frequency was observed.

A few reports describe the use of intact cells for transformation. Both incubation of cells with lithium acetate and particle bombardment (chapter 3) have been successfully used for the transformation of filamentous fungi (for references, see Table 1.1). These methods have the obvious advantage that the sometimes laborious protoplast preparation steps can be omitted.

Selection markers

Three types of selectable marker are used for selection of transformed cells: (a) a gene coding for a suppressor tRNA, (b) auxotrophic markers and (c) dominant selectable markers.

To date there is only one example of a suppressor tRNA gene (*su*-8, presumably a mutant tRNA gene) used as selection marker (Brygoo & Debuchy, 1985). Although this type of marker potentially can be used in each fungal strain which contains a suppressible chain termination mutation, no additional reports of the application of suppressor tRNA genes as selection marker have been published.

Auxotrophic markers are the most commonly used method for selection of transformants. Obviously, a prerequisite for their successful use is the presence of the appropriate mutation in the fungus. In Table 1.2 an overview is given of the auxotrophic markers which have been used. As can be seen from this Table, both homologous and heterologous markers can be used for transformation of fungi.

Some of the markers used (e.g. *pyrG*, *niaD* and *trpC*) have proved to be very useful, since they are functional in several species (Table 1.2). Furthermore, both *pyrG* and *niaD* are attractive markers for developing a gene transfer system in genetically poorly characterized fungal species, since the required mutants can be isolated by positive selection. In the case of *pyrG* they can be isolated by resistance against 5-fluoro-orotic acid (Van Hartingsveldt *et al.*, 1987; Goosen *et al.*, 1987) and in the case of *niaD* by resistance against chlorate (Unkles *et al.*, 1989). Since it is possible to select both for and against the mutant and wild-type phenotypes, these markers are also particularly useful for genetic manipulation strategies, such as gene-replacement experiments.

One of the obvious disadvantages of auxotrophic markers is the need to isolate a recipient strain with the appropriate mutation. With dominant selectable markers both wild-type and mutant strains can be transformed. A list of dominant markers which are utilized is given in Table 1.3. Several

Table 1.2. Auxotrophic selectable markers used for homologous and/or heterologous transformation of filamentous fungi.

Marker (species)**	Encoded function	Transformed species*	Reference
acuA$^+$ (*Ustilago maydis*)	acetyl-coA synthase	*Ustilago maydis*	Hargreaves & Turner (1989)
acuD$^+$ (*Aspergillus nidulans*)	isocitrate lyase	*Aspergillus nidulans*	Ballance & Turner (1986)
ade-2$^+$ (*Schizophyllum commune*)	unknown	*Phanerochaete chrysosporium*	Kornegay, Pribnow & Gold (1989)
am$^+$ (*Neurospora crassa*)	glutamate dehydrogenase	*Neurospora crassa*	Kinsey & Rambosek (1984)
amdS$^+$ (*Aspergillus nidulans*)	acetamidase	*Aspergillus nidulans*	Tilburn *et al.* (1983)
argB$^+$ (*Aspergillus nidulans*)	L-ornithine carbamoyl-transferase	*Aspergillus nidulans*	John & Peberdy (1984)
		Aspergillus niger	Buxton, Gwynne & Davies (1985)
inl$^+$ (*Neurospora crassa*)	unknown	*Neurospora crassa*	Akins & Lambowitz (1985)
leu$^+$ (*Mucor circinelloides*)	unknown	*Mucor circinelloides*	Van Heeswijck & Roncero (1984)
met$^+$ (*Aspergillus oryzae*)	unknown	*Aspergillus oryzae*	Iimura *et al.* (1987)
met-2$^+$ (*Ascobolus immersus*)	homoserine-O-trans acetylase	*Ascobolus immersus*	Goyon & Faugeron (1989)
niaD$^+$ (*Aspergillus nidulans*)	nitrate reductase	*Aspergillus niger*	Malardier *et al.* (1989)
		Fusarium oxysporum	Malardier *et al.* (1989)
niaD$^+$ (*Aspergillus niger*)	nitrate reductase	*Aspergillus niger*	Unkles *et al.* (1989)
		Penicillium chrysogenum	Whitehead *et al.* (1989)
niaD$^+$ (*Aspergillus oryzae*)	nitrate reductase	*Aspergillus oryzae*	Unkles *et al.* (1989)
		Aspergillus nidulans	Unkles *et al.* (1989)

Table 1.2. *continued.*

Marker (species)**	Encoded function	Transformed species*	Reference
nic-1⁺ (*Neurospora crassa*)	unknown	*Neurospora crassa*	Akins & Lambowitz (1985)
pkiA⁺ (*Aspergillus nidulans*)	pyruvate kinase	*Aspergillus nidulans*	De Graaff, Van den Broek & Visser (1988)
prn⁺ (*Aspergillus nidulans*)	proline catabolism	*Aspergillus nidulans*	Durrens *et al.* (1986)
pyr-3⁺ (*Ustilago maydis*)	dihydroorotase	*Ustilago maydis*	Banks & Taylor (1988)
pyr-4⁺ (*Neurospora crassa*)	orotidine-5'-phosphate decarboxylase	*Aspergillus nidulans*	Ballance, Buxton & Turner (1983)
pyr-6⁺ (*Ustilago maydis*)	orotidine-5'-phosphate decarboxylase	*Ustilago maydis*	Kronstad *et al.* (1989)
pyrG⁺ (*Aspergillus nidulans*)	orotidine-5'-phosphate decarboxylase	*Aspergillus nidulans*	Oakley *et al.* (1987)
pyrG/A⁺ (*Aspergillus niger*)	orotidine-5'-phosphate decarboxylase	*Aspergillus niger*	Van Hartingsveldt *et al.* (1987), Goosen *et al.* (1987)
		Aspergillus nidulans	Van Hartingsveldt *et al.* (1987)
pyrG⁺ (*Aspergillus oryzae*)	orotidine-5'-phosphate decarboxylase	*Aspergillus oryzae*	De Ruiter-Jacobs *et al.* (1989)
		Aspergillus niger	De Ruiter-Jacobs *et al.* (1989)
pyroA⁺ (*Aspergillus nidulans*)	unknown	*Aspergillus nidulans*	May *et al.* (1989)
qa-2⁺ (*Neurospora crassa*)	catabolic dehydroquinase	*Neurospora crassa*	Case *et al.* (1979)
QUTE⁺ (*Aspergillus nidulans*)	catabolic dehydroquinase	*Aspergillus nidulans*	Da Silva *et al.* (1986)
riboB⁺ (*Aspergillus nidulans*)	unknown	*Aspergillus nidulans*	Oakley *et al.* (1987)
trp-1⁺ (*Cochliobolus heterostrophus*)	trifunctional enzyme of tryptophan biosynthesis***	*Aspergillus nidulans*	Turgeon *et al.* (1986)

Table 1.2. *continued.*

Marker (species)**	Encoded function	Transformed species*	Reference
trp-1$^+$ (*Coprinus cinereus*)	tryptophan synthesis	*Coprinus cinereus*	Binninger *et al.* (1987)
trp-1$^+$ (*Schizophyllum commune*)	trifunctional enzyme of tryptophan biosynthesis***	*Schizophyllum commune*	Munoz-Rivas *et al.* (1986)
		Coprinus cinereus	Casselton & De La Fuenta Herce (1989)
trp-1$^+$ (*Neurospora crassa*)	trifunctional enzyme of tryptophan biosynthesis***	*Neurospora crassa*	Kim & Marzluf (1988)
trp-3$^+$ (*Neurospora crassa*)	tryptophan synthetase	*Neurospora crassa*	Vollmer & Yanofsky (1986)
trpC$^+$ (*Aspergillus nidulans*)	trifunctional enzyme of tryptophan biosynthesis***	*Aspergillus nidulans*	Yelton, Hamer & Timberlake (1984)
		Aspergillus niger	Goosen *et al.* (1989)
trpC$^+$ (*Aspergillus niger*)	trifunctional enzyme of tryptophan biosynthesis***	*Aspergillus nidulans*	Horng, Linz & Pestka (1989)
trpC$^+$ (*Phanerochaete chrysosporium*)	trifunctional enzyme of tryptophan biosynthesis***	*Coprinus cinereus*	Casselton & De La Fuente Herce (1989)
trpC$^+$ (*Penicillium chrysogenum*)	trifunctional enzyme of tryptophan biosynthesis***	*Penicillium chrysogenum*	Sánchez *et al.* (1987), Picknett *et al.* (1987)
		Aspergillus nidulans	Picknett *et al.* (1987)
ura-5$^+$ (*Podospora anserina*)	orotidylic acid pyrophosphorylase	*Podospora anserina*	Bégueret *et al.* (1984)

* listed here are the first species that have been transformed with the marker indicated by homologous or heterologous transformation. In several cases other species have subsequently been transformed with the same marker.

** the species from which the marker was isolated is indicated in parentheses.

*** encodes for glutamine amidotransferase, indoleglycerolphosphate synthetase and phosphoribosylanthranilate isomerase.

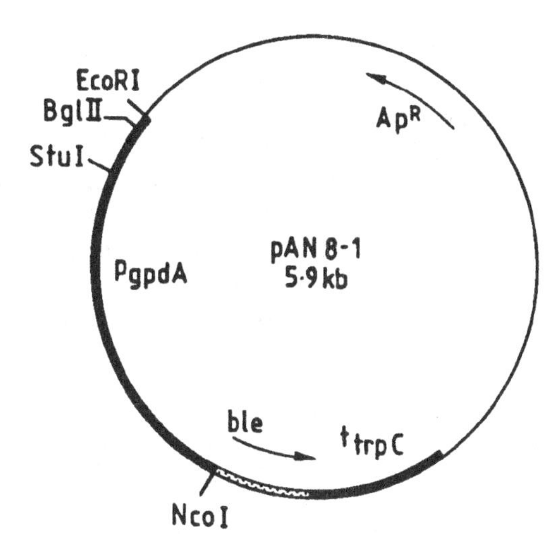

Fig. 1.1. Schematic representation of plasmid pAN8-1, which confers phleomycin resistance after transformation (Mattern, Punt & Van den Hondel, 1988). Thick line represents *A. nidulans* DNA, punctuated line *Streptoalloteichus hindustanus* DNA, and thin line *E. coli* DNA; p_{gpdA}, promoter region of the *gpdA* gene; t_{trpC}, terminator region of the *trpC* gene; *ble*, phleomycin resistance gene; Ap^R, ampicillin resistance gene. Arrows indicate the direction of transcription.

of these markers are 'broad-host range' markers which can be employed in different fungal species. All but one of these markers are based on drug-resistance. They consist of either mutant fungal genes such as benomyl resistant β-tubulin (*benA*, May *et al.*, 1985), or bacterial antibiotic-resistance genes provided with expression signals of filamentous fungi. The only exception is the acetamidase gene of *A. nidulans* (*amdS*, Kelly & Hynes, 1985), which is a nutritional marker. Transformants containing this gene are able to use acetamide or acrylamide as a sole nitrogen and carbon source. In general, fungi cannot readily use these compounds as such.

An example of a vector containing a bacterial resistance gene as selection marker is shown in Fig. 1.1. In this case the *Streptoalloteichus hindustanus* phleomycin resistance (*ble*) gene was introduced in a fungal expression vector containing the promoter region of the highly expressed *A. nidulans gpdA* gene and the terminator region of the *A. nidulans trpC* gene (Punt *et al.*, 1987). This vector and a similar one, pAN7-1, containing the *Escherichia coli* hygromycin B resistance gene have been used for the transformation of a wide variety of fungal species (Table 1.4).

Table 1.3. Dominant selectable markers

Marker*	Encoded function	Transformed species	Reference
amdS (*Aspergillus nidulans*)	acetamidase	*Aspergillus niger***	Kelly & Hynes (1985)
bar (*Streptomyces hygroscopicus*)	phosphinothricin acetylase	*Neurospora crassa*	Avalos *et al.* (1989)
benA (*Aspergillus nidulans*)	benomyl resistant β-tubulin	*Aspergillus nidulans*	May *et al.* (1985)
ble (*Escherichia coli*)	phleomycin binding protein	*Penicillium chrysogenum***	Kolar *et al.* (1988)
ble (*Streptoalloteichus hindustanus*)	phleomycin binding protein	*Aspergillus nidulans*/ *Aspergillus niger***	Mattern, Punt & Van den Hondel (1988)
5FI^r (*Coprinus cinereus*)	5-fluoroindole (feedback) resistant anthranilate synthetase	*Coprinus cinereus*	D. M. Burrows, T. J. Elliott & L. A. Casselton (unpublished)
G418^r (*Escherichia coli*)	geneticin/ neomycin/kanamycin phosphotransferase	*Ustilago maydis***	Banks (1983)
hph (*Escherichia coli*)	hygromycin B phosphotransferase	*Cephalosporium acremonium***	Queener *et al.* (1985)
oliC (*Aspergillus nidulans*)	mitochondrial ATP synthase subunit 9	*Aspergillus nidulans*	Ward, Wilkinson & Turner (1986)
oliC (*Aspergillus niger*)	mitochondrial ATP synthase subunit 9	*Aspergillus niger*	Ward *et al.* (1988)
oliC (*Penicillium chrysogenum*)	mitochondrial ATP synthase subunit 9	*Penicillium chrysogenum*	Bull, Smith & Turner (1988)
sul1 (*Escherichia coli*)	dihydropteroate synthetase	*Penicillium chrysogenum*	Carramolino *et al.* (1989)
tub (*Colletotrichum graminicola*)	benomyl resistant β-tubulin	*Colletotrichum graminicola*	Panaccione, McKierman & Hanau (1988)
tub-2 (*Neurospora crassa*)	benomyl resistant β-tubulin	*Neurospora crassa***	Orbach, Porro & Yanofsky (1986)
tubA (*Septoria nodorum*)	benomyl resistant β-tubulin	*Septoria nodorum***	Cooley & Caten (1989)

* the species from which the marker gene was isolated is indicated in parentheses. ** the species listed is the first species transformed with the marker. For the other markers transformation of only one species has been reported.

Types of vector

In general, vectors used for transformation experiments comprise *E. coli* plasmid DNA and the appropriate selectable marker. In most fungal species vector DNA becomes integrated into the genome of the host after transformation. Although considerable effort was undertaken to construct autonomously replicating vectors for *A. nidulans* and *Neurospora crassa*, using a strategy similar to that described for *Saccharomyces cerevisiae* (Stinchcomb, Struhl & Davis, 1979), no autonomous replication of the vector could be detected (Ballance & Turner, 1985; Buxton & Radford, 1984; Paietta & Marzluff, 1985; Van Gorcom, unpublished). In one case, however, a DNA sequence (the *A. nidulans ans1* sequence) which considerably enhances the transformation frequency was isolated. Nevertheless, even this vector did not replicate autonomously (Ballance & Turner, 1985).

For some other species, autonomously replicating vectors were successfully constructed by adding into an integration vector autonomously replicating sequences (ARS) (*Ustilago maydis*, Tsukuda *et al.*, 1988), the chromosomal ends of *Tetrahymena thermophila*, (*Podospora anserina*, Perrot, Barreau & Begueret, 1987), or the termini of naturally occurring linear plasmids of *Nectria haematococca* (*Ustilago maydis*, Samac & Leong, 1989).

In contrast to the results obtained for the ascomycetous fungi, *Neurospora* and *Aspergillus*, in zygomycetous fungi, like *Mucor circinelloides* (van Heeswijck, 1986), *Phycomyces blakesleeanus* (Revuelta & Jayaram, 1986), and *Absidia glauca* (Wostemeyer, Burmester & Weigel, 1987) autonomous replication of vectors was observed in most cases. Autonomous replication was also observed for a filamentous yeast species, *Trichosporon cutaneum* (Glumoff *et al.*, 1989) transformed with pAN7-1 (see above).

Fate of transforming DNA

As already mentioned, in most filamentous fungi vector DNA is integrated into the genome. Biochemical analysis of the DNA of transformants indicates that when a homologous selection marker is used, in general three types of integration events can occur: type I, integration of the vector by homologous recombination; type II, ectopic integration of the vector (or vector sequences) by non-homologous recombination; and type III, gene replacement. For most homologous selectable markers, predominantly homologous interactions (type I and III integrations) occur. However, in some cases type II transformants are preferentially found, e.g. in *A. nidulans* with the *amdS* gene (Wernars *et al.*, 1985) or the *prn* gene cluster (Durrens *et al.*, 1986), and in *Ascobolus immersus* with the *met2* gene (Goyon & Faugeron, 1989). Also, in *Coprinus cinereus*

Table 1.4. Fungal species successfully transformed with the vectors pAN7-1 and/or pAN8-1

Transformed species	Vector		Reference
	pAN7-1	pAN8-1	
Acremonium chrysogenum	+	ND	A. W. Smith, M. Ramsden & J. F. Peberdy (unpubl.)
Aspergillus nidulans	+	+	Punt *et al.* (1987)
Aspergillus niger	+	+	Punt *et al.* (1987)
Aspergillus ficuum	+	ND	Mullaney, Punt & Van den Hondel (1988)
Aspergillus oryzae	–	+	Mattern, Punt & Van den Hondel (1988)
Aspergillus giganteus	+	ND	Wnendt, Jacobs & Stahl (1990)
Claviceps purpurea	+	ND	Comino *et al.* (1989)
Cryphonectria parasitica	+	ND	Churchill *et al.* (1990)
Curvularia lunata	+	ND	Osiewacz & Weber (1989)
Fulvia fulvum	+	+	Oliver *et al.* (1987)
Fusarium culmorum	+	ND	H. Curragh, R. Marchant, H. Mooibroek & J. G. H. Wessels (unpubl.)
Leptosphaeria maculans	+	ND	Farman & Oliver (1988)

predominantly type II transformants are observed when the *TRP-1* marker is used (Binninger *et al.*, 1987). Transformation of *Ascobolus immersus* with vector DNA linearised by cutting within the marker sequence or with circular single-stranded vector DNA preferentially results in type I integration events (Goyon & Faugeron, 1989).

In the case of heterologous selectable markers integration will always occur through non-homologous recombination, seemingly at random sites in the genome.

Genetic manipulation

The availability of different gene transfer systems with different characteristics permits a molecular-genetic study of all kinds of biologically

Table 1.4. *continued.*

Transformed species	Vector		Reference
	pAN7-1	pAN8-1	
Neurospora crassa	+	ND	Staben *et al.* (1989)
Penicillium chrysogenum	–	+	Kolar *et al.* (1988)
Penicillium roquefortii	+	ND	N. Durand, P. Reymond & M. Fevre (unpubl.)
Pseudocercosporella herpotrichoides	+	ND	Blakemore *et al.* (1989)
Schizophyllum commune	+	ND	Mooibroek *et al.* (1990)
Septoria nodorum	+	ND	Cooley *et al.* (1988)
Talaromyces emersonii	ND	+	S. Jain, H. Durand & G. Tiraby (unpubl.)
Trichoderma harzianum	+	ND	Goldman, Van Montagu & Herrera-Estrella (1990); C. J. Ulhoa, M. H. Vainstein & J. F. Peberdy (unpubl.)
Trichoderma hamatum	+	ND	C. J. Ulhoa, M. H. Vainstein & J. F. Peberdy (unpubl.)
Trichoderma viride	+	ND	Herrera-Estrella, Goldman & Van Montagu (1990)
Trichosporon cutaneum	+	+	Glumoff *et al.* (1989)

interesting processes by isolation, characterisation and functional analysis of the genes and gene products involved. To perform these studies, specific vectors are constructed which facilitate genetic manipulation such as cloning of a gene by complementation of a mutation, gene disruption or gene replacement, and analysis of expression signals *in vivo*.

To illustrate the possibilities of genetic manipulation for molecular genetic studies examples will be given of research on *Aspergillus* that is in progress in our laboratory. The first example concerns experiments that have been performed to prove that a gene encoding a functional benzoate-*p*-hydroxylase gene of *Aspergillus niger* was cloned. In the second example, the application of different expression-analysis vectors for a functional

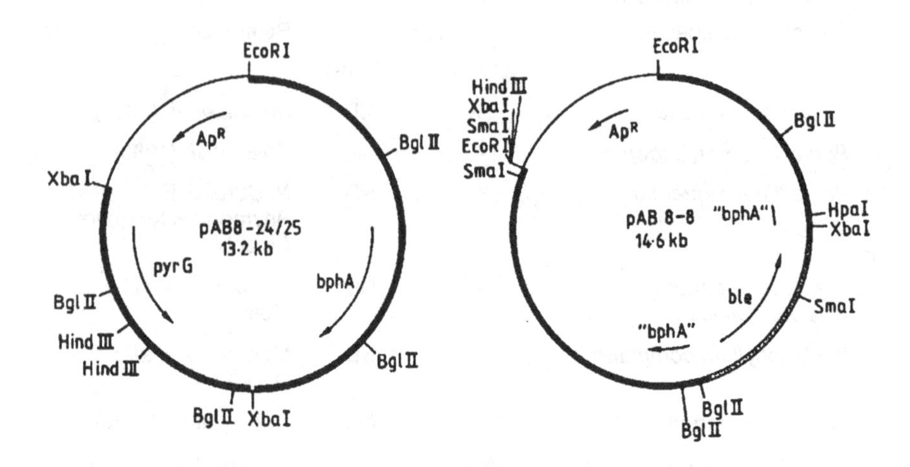

Fig. 1.2. Schematic representation of plasmid pAB8-8, which contains the disrupted *A. niger bphA* gene and the plasmids pAB8-24 and pAB8-25, which contain the *bphA* gene and, respectively, the wildtype or a mutant allele of the *pyrG* gene (Van Gorcom & Van den Hondel, 1988) of *A. niger* respectively. The disrupted *bphA* gene was obtained by replacing an *Eco*RV segment, located within the *bphA* gene, with the phleomycin resistance unit of pAN8-1. Thick line represents *A. niger* DNA, punctuated line *Streptoalloteichus hindustanus* DNA and thin line *E. coli* DNA; *ble*, phleomycin resistance gene; Ap^R, ampicillin resistance gene; '*bphA*', 5'- or 3'-terminal part of the *bphA* gene. Arrows indicate the direction of transcription.

analysis of the promoter region of the *Aspergillus* genes *gpdA*, *niaD* and *niiA* will be described. The third example deals with a study of the influence of different signal sequences on the efficiency of production of prochymosin in *A. niger*.

Cloning of a functional bphA *gene of* A. niger

Benzoate is metabolized by *A. niger* in a series of steps of which the first is *p*-hydroxylation of the aromatic ring of benzoate, carried out by benzoate-*p*-hydroxylase (BPH). Several mutants, disturbed in BPH activity, have been isolated (Boschloo & Bos, in preparation). These mutations were shown to belong to one complementation group, therefore the mutation was named *bphA*.

A cosmid clone, pAB8-1, containing the putative *bphA* gene, was isolated by differential hybridiaztion techniques. The gene was localized on a 6·2 kb *Eco*RI–*Pvu*II fragment, which was subcloned in pUC19, resulting in pAB8-22 (Van Gorcom *et al.*, 1990). Introduction of pAB8-1 or pAB8-22 DNA into an *A. niger bphA* mutant resulted in the restoration

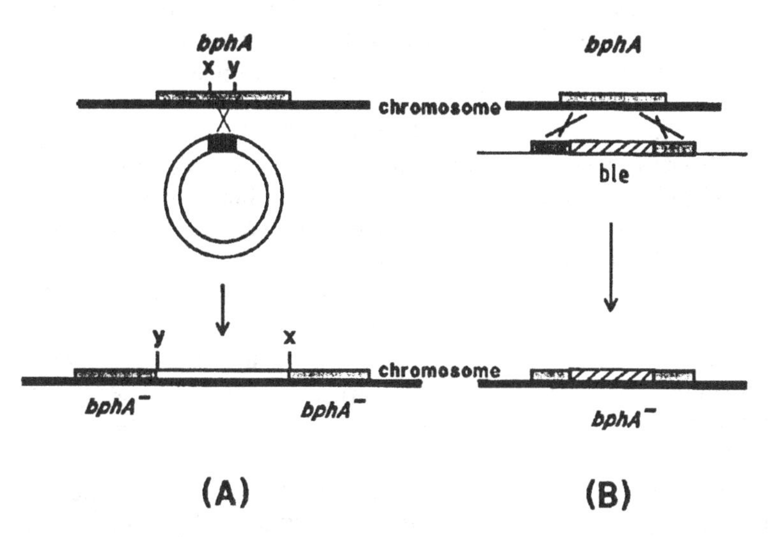

Fig. 1.3. Strategies to disrupt the *bphA* gene. Thin lines represent plasmid DNA and thick line chromosomal DNA. Shaded boxes represent the *bphA* gene or part of the gene. Hatched boxes represent the phleomycin resistance unit. Two restriction sites within the *bphA* gene are indicated with X and Y. (A) Disruption of the *bphA* gene by transformation with a plasmid which contains an internal restriction fragment. The recombination event shown results in formation of a duplication of *bphA* with the leftward copy lacking the 3' end of the gene and the rightward copy lacking the 5' end. (B) Disruption of the *bphA* gene by transformation with a linear fragment which contains a mutant allele of the *bphA* gene obtained by replacing an internal fragment with the phleomycin resistance unit. Recombination between the rightward and leftward homologous regions of the DNA fragment and the corresponding chromosomal regions results in a gene replacement of the wildtype gene with the mutant (disrupted) *bphA* allele.

---

of the ability to grow on benzoate, suggesting that the DNA fragment contained the *bphA* gene.

Although remote, it cannot be completely excluded that a suppressor of the *bphA* mutation had been cloned. One approach to exclude this possibility is to disrupt the cloned gene, replace the chromosomal gene by the disrupted equivalent and test for the inability to grow on benzoate.

Two methods regularly used in gene-disruption experiments are indicated in Fig. 1.3. In both cases the disruption vector contains a non-functional copy of the chromosomal gene to be disrupted. The method indicated in Fig. 1.3A requires knowledge about the exact position of the gene in the cloned fragment, whereas for the method indicated in Fig. 1.3B this is not necessary. To obtain an *A. niger* strain in which the *bphA* gene was disrupted, the method indicated in Fig. 1.3B was chosen. For the disruption experiment, plasmid pAB8-8 was constructed (Fig. 1.2) which contains the non-functional *bphA* gene. In this plasmid part of the *bphA* sequences has been replaced by the phleomycin resistance unit of pAN8-1 (Mattern, Punt & Van den Hondel, 1988). Transformation of *A. niger* wild type with the isolated *Eco*RI fragment of pAB8-8 resulted in a number of phleomycin resistant colonies. Southern blot analysis revealed that in about 10% of the transformants a gene replacement had occurred. Further analysis showed that these transformants were not able to grow on benzoate as carbon source. This result confirms that the *bphA* gene and not a suppressor gene had been cloned.

Further evidence for cloning of the benzoate-*p*-hydroxylate-encoding gene was obtained from the DNA sequence of the *bphA* gene. Sequence comparison showed that the *bphA* gene encoded a cytochrome P450 mono-oxygenase, as might be expected.

Another important issue was the question whether the cloned gene was a functional copy of the *bphA* gene. To answer this question it was necessary to prove that the *bphA* mutation was complemented by the product of the cloned gene. Therefore an *A. niger bph* strain was transformed with a plasmid containing the cloned gene and transformants were isolated in which the plasmid was integrated at an ectopic locus. Growth of these transformants on benzoate would indicate that a functional gene had been cloned. To achieve ectopic integration, the *A. niger pyrG* selection marker was cloned into pAB8-22 resulting in plasmid pAB8-24 (Fig. 1.2). Van Hartingsveldt *et al.* (1987) previously had found that a vector containing this selection marker is integrated at the *pyrG* locus in about 50% of *A. niger* transformants. However, Southern analysis of 48 transformants, obtained with pAB8.24, revealed that none of these transformants contained a vector integrated at the *pyrG* locus. Further analysis indicated that in most transformants the vector was integrated at the *bphA* locus.

To overcome the problem of preferential integration at the *bphA* locus, a mutant allele of the *A. niger pyrG* gene (Van Gorcom & Van den Hondel, 1988) was cloned in pAB8-22, resulting in pAB8-25 (Fig. 1.2). This mutant allele was constructed by introduction of a frameshift mutation which inactivates the marker gene. Transformation with the mutant allele as selection marker can result in Pyr^+ transformants only through type I or type III integration events. Analysis by Southern blotting of transformants obtained with pAB8-25 revealed that 14 out of 32 contained a single copy of this plasmid integrated at the *pyrG* locus. These transformants also showed a restored ability to grow on benzoate, indicating that, indeed, a functional *bphA* gene had been cloned. As demonstrated by Southern analysis the other transformants resulted from a gene replacement at the *pyrG* locus. As expected, these transformants could not grow on benzoate.

Vectors for analysis of expression signals from Aspergillus *genes*

In both fundamental and applied molecular biological research on filamentous fungi the unravelling of the mechanism of gene expression is a very important topic. Interesting biological processes, such as development, differentiation and carbon and nitrogen metabolism are regulated at the level of gene expression. A wealth of classical genetic information is available for these processes, but, until recently, hardly any molecular genetic research was carried out. To provide an easy way to assay the expression and regulation of various genes, we developed reporter vectors for filamentous fungi (Van Gorcom *et al.*, 1986; Van Gorcom & Van den Hondel, 1988; Roberts *et al.*, 1989). In these vectors the analysis of fungal expression signals can be carried out by fusion of these signals to the *E. coli* reporter genes, *lacZ* or *uidA* encoding β-galactosidase and β-glucuronidase, respectively. The products of these genes can be assayed both qualitatively and quantitatively with easy and sensitive methods. For proper analysis of expression signals, it is essential that integration of one copy of the expression unit can be achieved at a specific location on the chromosome of the recipient. To fulfil this requirement, homologous selection markers were introduced in these vectors. An even higher (relative) frequency of homologous integration could be obtained by using mutant selection markers. These mutant selection markers were constructed by introduction of a frameshift mutation which inactivates the marker gene. Thus, only intragenic recombination (Type I or III integration) between the mutant selection marker on the vector and the mutant allele in the genome will result in prototrophic transformants. Although the transformation frequency obtained with this type of marker is much reduced (about 10-100 fold), Southern analysis of only a few transformants is sufficient to identify transformants with a single copy at the locus chosen (Table 1.5). Also, linearisation of the vector with a restriction enzyme which cuts in the marker gene, increases the relative frequency of Type I integration (Table 1.5).

The promoters of the *gpdA* genes of both *A. niger* and *A. nidulans* were studied in *A. niger* with the use of one of these vectors (Fig. 1.4). Single copy transformants, obtained with the two p_{gpdA}-*lacZ* fusion constructs, were assayed for β-galactosidase activity. In both cases efficient β-galactosidase expression was obtained (Table 1.6), whereas in untransformed strains or strains transformed with pAB94-12 (vector without promoter sequences inserted) no significant β-galactosidase activity was detected. From these results we concluded that the promoter of the *gpdA* gene of

Table 1.5. Results of Southern analysis of *A. nidulans* transformants obtained with pAN5-d1 and derivatives

Vector[1]	Transformation frequency[2]	Percentage of LacZ$^+$ transformants[3]	Type of integration[4]		
			A	B	C
pAN5.d1	20-40	60%	0/19	4/19	15/19
pAN5.d1$_{(BglII\ digest)}$	40-100	90%	1/10	5/10	4/10
pAN5.d1$_{BglII}$	0·1-1	40%	5/10	3/10	2/10

[1] Vector pAN5-d1 contains a p_{gpdA}-*LacZ* fusion and the wildtype *argB* gene as selection marker for *Aspergillus* transformation (Punt *et al.*, 1990). The vector contains a unique *Bgl*II site in the coding region of the *argB* gene. Analysis of the transformants obtained with pAN5-d1, with a *Bgl*II digest of pAN5-d1 and with pAN5-d1$_{BglII}$, in which the unique *Bgl*II site was filled in with PolIK resulting in a frame shift mutation in the *argB* gene (Punt *et al.*, 1990), was carried out. Vectors were introduced into *A. nidulans* ArgB$^-$ (*methG2, biA1, argB2*).

[2] Transformation frequency is given as transformants per μg of vector DNA.

[3] The percentage of LacZ$^+$ transformants was determined by plating transformants on agar plates containing XGal (van Gorcom *et al.*, 1985). In all cases both LacZ$^+$ and LacZ$^-$ transformants were observed. The LacZ$^-$ transformants probably arose from gene replacement events.

[4] Southern analysis of a number of LacZ$^+$ transformants was carried out. The transformants were classified in three categories; A, single copy integration of the vector at the *argB* locus; B, multiple copy (tandem) integration of vector molecules at the *argB* locus; C, ectopic integration, in some cases in combination with single or multiple copy homologous integration.

A. nidulans and *A. niger* are both very efficient in *A. niger*. Further analysis of the organisation of the expression signals of the *A. nidulans* gene *gpdA* with similar vectors developed for *A. nidulans* is in progress in our laboratory (Punt *et al.*, 1990).

Recent research has shown that many fungal genes involved in developmental and metabolic pathways are organised as gene clusters (Gurr, Unkles & Kinghorn, 1988). Frequently, these clustered genes are coordinately expressed from divergently transcribing intergenic promoter regions. For the analysis of such intergenic regions a twin reporter vector was developed (Fig. 1.5). The usefulness of this vector can be inferred from the functional analysis of the intergenic region between the *A. nidulans* nitrate reductase (*niaD*) and nitrite reductase (*niiA*) genes. As shown in Table 1.7, both nitrate induction and nitrogen metabolite (ammonium) repression is observed for the reporter genes. Thus, the

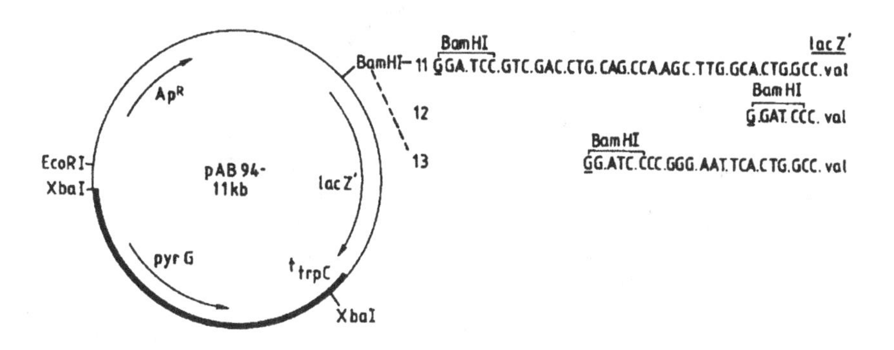

Fig. 1.4. Schematic representation of expression analysis vectors pAB94-11 to 13 for *A. niger* (Van Gorcom & Van den Hondel, 1988). The different vectors contain a unique *Bam*HI site in one of the three reading frames in front of the *lacZ'* gene (the protein coding region of the *E. coli lacZ* gene lacking the first eight codons). Thick line represents *A. niger* DNA (*Xba* I fragment) and *A. nidulans* DNA. Thin line represents *E. coli* DNA; t_{trpC}, terminator region of the *trpC* gene; Ap^R, ampicillin resistance gene; *pyrG*, mutant allele of the *A. niger pyrG* gene. Arrows indicate the direction of transcription.

Table 1.6. β-Galactosidase expression in *A. niger* transformants containing p_{gpdA}-*lacZ* fusion genes

Strain[1]	p_{gpdA}	βGAL activity[2]
AB4-1 [pAB94-53] 4	*A. niger*	8570
6		8380
7		7770
AB4-1 [pAB94-121] 4	*A. nidulans*	5160
13		5480
17		5350
AB4-1	–	<10

[1] Vectors pAB94-53 and pAB94-121, derivatives of pAB94-11/12/13, containing the promoter region of the *gpdA* gene of *A. niger* and *A. nidulans*, respectively, fused to the *LacZ* gene, were introduced into *A. niger* AB4-1 (*cspA*1, *pyrG*). Transformants with a single copy of the vector integrated at the *pyrG* locus were identified by Southern analysis.

[2] Enzyme activity is given in units (mg protein)$^{-1}$ and was measured as described by Van Gorcom *et al.* (1985).

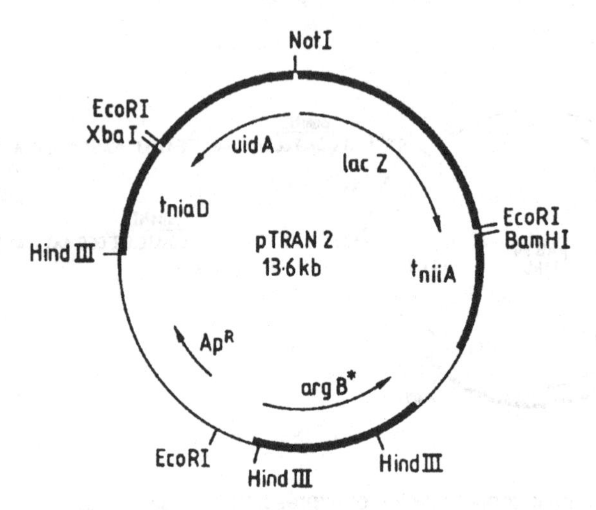

Fig. 1.5. Schematic representation of the twin reporter vector pTRAN2. Thin line represent pBR322 DNA; thick lines, *A. nidulans* DNA and *E. coli* DNA (*Eco*RI fragment) which contains the coding region of the *E. coli* genes *lacZ* and *uidA* both without translation initiation codon. A unique *Not*I site is placed between these genes; t_{niaD}, terminator region of the *niaD* gene; t_{niiA}, terminator region of the *niiA* gene; Ap^R, ampicillin resistance gene; $argB^+$, mutant allele of the *A. nidulans argB* gene containing a frameshift mutation. Arrows indicate the direction of transcription.

expression of the reporter genes *lacZ* and *uidA* faithfully represents the regulated gene expression of the genes *niaD* and *niiA* (Cove, 1979).

Expression of prochymosin

Several filamentous fungi are able to produce large amounts of extracellular proteins. Due to this property, several groups, including ours, are carrying out research to evaluate the potential of these strains for the production of heterologous proteins. One of the questions we addressed in our research on expression and secretion of heterologous, extracellular proteins in *A. niger* is the influence of different signal sequences on the efficiency of protein production/secretion. To answer this question experiments were performed to analyze the production of prochymosin with four different gene fusions (van Hartingsveldt *et al.*, 1990). These fusions were placed under the control of the expression signals of the *A. niger* glucoamylase (*glaA*) gene. To facilitate proper comparison, transformants containing a single copy of the expression unit integrated at the *glaA* locus were isolated. For this purpose four different prochymosin expression vectors, pAB64-72 to pAB64-75 (Fig. 1.6), were used. Transformation of *A. niger* with *Hin*dIII-linearised pAB64-72 to 75 resulted in a number of hygromycin B resistant transformants. Southern analysis demonstrated

Table 1.7. Expression of the reporter genes in pTRAN2-1A transformants

	Relative enzyme activities[2]			
Strain[1]:	G324		SAA1012	
	βGUS	βGAL	βGUS	βGAL
proline	10	20	320	180
nitrate+proline	100	100	490	260
nitrate+ammonium	20	40	10	20
ammonium	2	10	2	4

[1] Vector pTRAN2-1A, a derivative of pTRAN2 (Fig. 1.4), containing the *A. nidulans niaD–niiA* intergenic promoter region was introduced in *A. nidulans* strains G324 (*wA*3, *yA*2, *methH*2, *argB*2, *galA*1, *sC*12, *ivoA*1) and SAA1012 (*fwA*1, *yA*2, *methH*2, *pabaA*1, *argB*2, *niiA-niaD* Δ509). Single copy transformants were identified by Southern analysis. In all cases the β-glucuronidase (βGUS) expression is a result of the activity of the expression signals of the *niaD* gene, and the β-galactosidase (βGAL) expression results from *niiA* gene expression signals.

[2] Mycelial extracts were prepared from cells cultivated for 16 to 18 h in minimal growth medium with appropriate supplements and 10 mM of the indicated nitrogen sources. The enzyme activities were determined as described previously (Van Gorcom *et al.*, 1985; Roberts *et al.*, 1989) and are expressed relative to the activities of the G324[pTRAN2-1A] transformants induced with nitrate (= 100). In a representative experiment specific activities of 80 nmol *p*-nitrophenol min^{-1} (mg protein)$^{-1}$ for βGUS and 310 nmol *o*-nitrophenol min^{-1} (mg protein)$^{-1}$ for βGAL were found for the G324[pTRAN2-1A] transformants induced with nitrate.

Table 1.8. Analysis of prochymosin production in *A. niger*

Strain[1]	Signal peptide	Western[2] ($\mu g\ ml^{-1}$)	MCA[2] ($U\ ml^{-1}$)
AB64-72	signal sequence of prochymosin	6·2	8·6
AB64-73	signal sequence of *glaA*	11·3	19·5
AB64-74	signal sequence of *glaA* + 6 additional amino acids	4·1	3·2
AB74-75	signal sequence of *glaA* + 53 additional amino acids	10·2	19·1

[1] Vectors pAB64-72 to 75, linearised by cutting with *Hind*III, were introduced into *A. niger*. Transformants in which the *glaA* gene is replaced by the prochymosin fusion-genes, were identified by Southern hybridization (Van Hartingsveldt *et al.*, 1990).

[2] Medium samples from cells cultivated for 24 h in induction medium, were analyzed for the presence of prochymosin by Western blotting (Western) and milk clotting assay (MCA) (Van Hartingsveldt *et al.*, 1990).

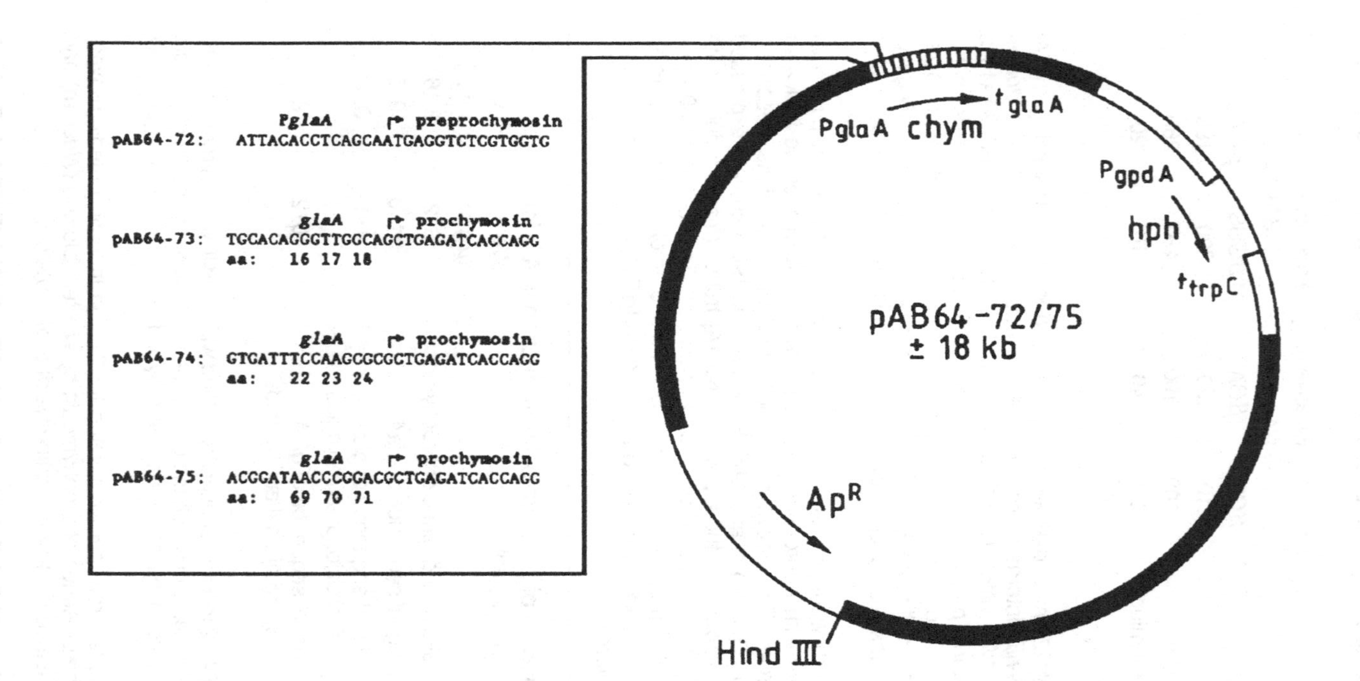

Fig. 1.6. Schematic representation of four different prochymosin-expression vectors pAB64-72 to 75 (Van Hartingsveldt *et al.*, 1990). Sequences at the glucoamylase-preprochymosin junction are indicated at the lefthand side. Thick lines represent *A. niger* DNA from the *glaA* locus, open lines *A. nidulans* DNA, hatched lines preprochymosin DNA and thin lines *E. coli* DNA; p_{gpdA}, promoter region of the *gpdA* gene; p_{glaA}, promoter region of the *glaA* gene; t_{trpC}, terminator region of the *trpC* gene; t_{glaA}, terminator region of the *glaA* gene; *hph*, hygromycin resistance gene; Ap^R, ampicillin resistance gene; *chym*, calf preprochymosin. Arrows indicate the direction of transcription.

that in about 10% of these transformants the resident *glaA* gene was replaced by the expression/secretion unit. Similar results were obtained with circular vector DNA, though with a three- to five-fold lower frequency.

Transformants which contained one copy of the expression/secretion unit were analyzed for prochymosin production 24 h after induction of the *glaA* promoter with starch (Van Hartingsveldt *et al.*, 1990). As shown in Table 1.8 similar levels of prochymosin were produced with pAB64-73 (18 amino acids of *glaA*) and pAB64-75 (71 amino acids of *glaA*). With 24 amino acids of *glaA* in front of prochymosin, or with the signal sequence of prochymosin itself, a lower production level was observed.

Although the reasons for the observed differences are obscure, our results clearly demonstrate that gene fusions containing different 5' sequences influence the production level of prochymosin.

Conclusions

During the last few years the development of gene transfer systems has been described for more than fifty fungal species. Transformation of most species could be achieved with heterologous auxotrophic markers or dominant selectable markers. Usually the marker gene is expressed from fungal, mainly *A. nidulans*, expression signals which were shown to be functional in most fungal species.

A number of strategies are now available for the development of gene transfer systems for hitherto poorly characterized fungal species. As illustrated in the first part of this chapter, these strategies comprise the following aspects. Firstly, methods for the introduction of vector DNA (Table 1.1). Secondly, a large number of auxotrophic and dominant selectable markers (Tables 1.2 to 1.4). Thirdly, efficient strategies for the isolation of auxotrophic mutant strains.

The main purpose for the development of gene transfer systems is application of these systems for molecular genetic studies. In the second part of this chapter several applications were illustrated with examples taken from research carried out in our laboratory. Genetic manipulation experiments were carried out (a) to disrupt the *bphA* gene of *A. niger*; (b) to analyze expression signals in *A. nidulans* and *A. niger*; (c) to direct expression-analysis vectors at specific sites of the genome such as the *argB* locus of *A. nidulans* or the *pyrG* locus of *A. niger* and (d) to perform gene replacement experiments in which the *glaA* gene of *A. niger* was replaced by chimeric prochymosin genes. These examples, as well as others described in the recent literature, indicate that most strategies and tools for genetic manipulation in filamentous fungi are now available, especially for *A. nidulans*, *A. niger* and *Neurospora crassa*. Extensive molecular genetic studies of many interesting biological processes occurring in filamentous fungi can now be carried out using these strategies and tools.

Acknowledgements We wish to thank Peter Pouwels for critical reading of the manuscript and fruitful discussions. Gerry van Oossanen is acknowledged for typing the manuscript. Part of the work described in this chapter was financially supported by Gist-brocades, Delft, The Netherlands, DSM research, Geleen, The Netherlands and training grants from the EC.

References

Akins, R. A. & Lambowitz, A. M. (1985). General method for cloning *Neurospora crassa* nuclear genes by complementation of mutants. *Molecular and Cellular Biology* **5**, 2272-2278.

Armaleo, D., Ye, G. -N., Klein, T. M., Shark, K. B., Sanford, J. C. & Johnston, S. A. (1990). Biolistic nuclear transformation of *Saccharomyces cerevisiae* and other fungi. *Current Genetics* **17**, 97-103.

Avalos, J., Geever, R. F. & Case, M. E. (1989). Bialaphos resistance as a dominant selectable marker in *Neurospora crassa*. *Current Genetics* **16**, 369-372.

Ballance, D. J., Buxton, F. P. & Turner, G. (1983). Transformation of *Aspergillus nidulans* by the orotidine-5'-phosphate decarboxylase gene of *Neurospora crassa*. *Biochemical and Biophysical Research Communications* **112**, 284-289.

Ballance, D. J. & Turner, G. (1985). Development of a high frequency transforming vector for *Aspergillus nidulans*. *Gene* **36**, 321-331.

Ballance, D. J. & Turner, G. (1986). Gene cloning in *Aspergillus nidulans*: isolation of the isocitrate lyase gene (*acuD*). *Molecular and General Genetics* **202**, 271-275.

Banks, G. R. (1983). Transformation of *Ustilago maydis* by a plasmid containing yeast 2 micron DNA. *Current Genetics* **7**, 73-77.

Banks, G. R. & Taylor, S. Y. (1988). Cloning of the *PYR3* gene of *Ustilago maydis* and its use in DNA transformation. *Molecular and Cellular Biology* **8**, 5417-5424.

Bégueret, J., Razanamparany, V., Perrot, M. & Barreau, C. (1984). Cloning gene *ura5* for the orotidylic acid pyrophosphorylase of the filamentous fungus *Podospora anserina*: transformation of protoplasts. *Gene* **32**, 487-492.

Bej, A. K. & Perlin, M. H. (1989). A high efficiency transformation system for the basidiomycete *Ustilago maydis* employing hygromycin resistance and lithium-acetate treatment. *Gene* **80**, 171-176.

Binninger, O. M., Skrzynia, C., Pukkila, P. J. & Casselton, L. A. (1987). DNA-mediated transformation of the basidiomycete *Coprinus cinereus*. *EMBO Journal* **6**, 835-840.

Blakemore, E. J. A., Dobson, M. J., Hocart, M. J., Lucas, J. A. & Peberdy, J. F. (1989). Transformation of *Pseudocercosporella herpotrichoides* using heterologous genes. *Current Genetics* **16**, 177-180.

Brygoo, Y. & Debuchy, R. (1985). Transformation by integration in *Podospora anserina*.I. Methodology and phenomenology. *Molecular and General Genetics* **200**, 128-131.

Bull, J. H., Smith, D. J. & Turner, G. (1988). Transformation of *Penicillium chrysogenum* with a dominant selectable marker. *Current Genetics* **13**, 377-382.

Buxton, F. P. & Radford, A. (1984). The transformation of mycelial spheroplasts of *Neurospora crassa* and the attempted isolation of an autonomous replicator. *Molecular and General Genetics* **196**, 339-344.

Buxton, F. P., Gwynne, D. I. & Davies, R. W. (1985). Transformation of *Aspergillus niger* using the *argB* gene of *Aspergillus nidulans*. *Gene* **37**, 207-214.

Carramolino, L., Lozano, M., Pérez-Aranda, A., Rubio, V. & Sanchez, E. (1989). Transformation of *Penicillium chrysogenum* to sulfonamide resistance. *Gene* **77**, 31-38.

Case, M. E., Schweizer, M., Kushner, S. R. & Giles, N. H. (1979). Efficient transformation of *Neurospora crassa* by utilizing hybrid plasmid DNA. *Proceedings of the National Academy of Sciences, USA* **76**, 5259-5263.

Casselton, L. A. & De La Fuente Herce, A. (1989). Heterologous gene expression in the basidiomycete fungus *Coprinus cinereus*. *Current Genetics* **16**, 35-40.

Churchill, A. C. L., Guffetti, L. M., Hansen, D. R., Van Etten, H. D. & Van Alfen, N. K. (1990). Transformation of the fungal pathogen *Cryphonectria parasitica* with a variety of heterologous plasmids. *Current Genetics* **17**, 25-31.

Comino, A., Kolar, M., Schwab, H. & Socic, H. (1989). Heterologous transformation of *Claviceps purpurea*. *Biotechnology Letters* **11**, 389-392.

Cooley, R. N., Shaw, R. K, Franklin, F. C. H. & Caten, C. E. (1988). Transformation of the phytopathogenic fungus *Septoria nodorum* to hygromycin B resistance. *Current Genetics* **13**, 383-390.

Cooley, R. N. & Caten, C. E. (1989). Cloning and characterization of the β-tubulin gene and determination of benomyl resistance in *Septoria nodorum*. In *Proceedings, EMBO Alko Workshop on Molecular Biology of Filamentous Fungi*, vol. 6, (ed. H. Nevailainen & M. Penttilä), pp. 207-216. Foundation of Biotechnical and Industrial Fermentation Research: Helsinki.

Cove, D. J. (1979). Genetic studies of nitrate assimilation in *Aspergillus nidulans*. *Biological Reviews* **54**, 291-327.

Da Silva, A. J. F., Whittington, H., Clements, J., Roberts, C. & Hawkins, A. R. (1986). Sequence analysis and transformation by the catabolic 3-dehydroquinase (*QUTE*) gene from *Aspergillus nidulans*. *Biochemical Journal* **240**, 481-488.

De Graaff, L., Van den Broek, H. & Visser, J. (1988). Isolation and transformation of the pyruvate kinase gene of *Aspergillus nidulans*. *Current Genetics* **13**, 315-321.

De Ruiter-Jacobs, Y. M. J. T., Broekhuijsen, M., Campbell, E. I., Unkles, S. E., Kinghorn, J. R., Contreras, R., Pouwels, P. H. & van den Hondel, C. A. M. J. J. (1989). A gene transfer system based on the homologous *pyrG* gene and efficient expression of bacterial genes in *Aspergillus oryzae*. *Current Genetics* **16**, 159-163.

Durrens, P., Green, P. M., Arst, H. N. & Scazzocchio, C. (1986). Heterologous insertion of transforming DNA and generation of new deletions associated with transformation in *Aspergillus nidulans*. *Molecular and General Genetics* **203**, 544-549.

Farman, M. L. & Oliver, R. P. (1988). The transformation of protoplasts of *Leptosphaeria maculans* to hygromycin B resistance. *Current Genetics* **13**, 327-330.

Fincham, J. R. S. (1989). Transformation in fungi. *Microbiological Reviews* **53**, 148-170.

Glumoff, V., Käppeli, O., Fiechken, A. & Reiser, J. (1989). Genetic transformation of the filamentous yeast *Trichosporon cutaneum* using dominant selection markers. *Gene* **84**, 311-318.

Goldman, G. H., Van Montagu, M. & Herrera-Estrella, A. (1990). Transformation of *Trichoderma harzianum* by high-voltage electric pulse. *Current Genetics* **17**, 169-174.

Goosen, T., Bloemheuvel, G., Gysler, C., De Bie, D. A., Van den Broek, H. W. J. & Swart, K. (1987). Transformation of *Aspergillus niger* using the homologous orotidine-5'-phosphate-decarboxylase gene. *Current Genetics* **11**, 499-503.

Goosen, T., van Engelenburg, F., Debets, F., Swart, K., Bos, K. & Van den Broek, H. (1989). Tryptophan auxotrophic mutants in *Aspergillus niger*. Inactivation of the *trpC* gene by cotransformation mutagenesis. *Molecular and General Genetics* **219**, 282-288.

Goosen, T., Bos, C. J. & Van den Broek, H. W. J. (1991). Transformation and gene manipulation in filamentous fungi: an overview. In *Handbook of Applied Mycology* (Fungal Biotechnology volume 4), (ed. D. K. Arora, K. G. Mukerji & R. P. Elander), in press. M. Dekker: New York.

Goyon, C. & Faugeron, G. (1989). Targeted transformation of *Ascobolus immersus* and *de novo* methylation of the resulting duplicated DNA sequences. *Molecular and Cellular Biology* **9**, 2818-2827.

Gurr, S. J., Unkles, S. E. & Kinghorn, J. R. (1988). The structure and organisation of nuclear genes of filamentous fungi. In *Gene Structure in Eukaryotic Microbes*, Society of General Microbiology Special Publication volume 23, (ed. J. R. Kinghorn), pp. 93-139. IRL Press: Oxford.

Hargreaves, J. A. & Turner, G. (1989). Isolation of the acetyl-CoA synthase gene from the corn smut pathogen, *Ustilago maydis*. *Journal of General Microbiology* **135**, 2675-2678.

Henson, J. M., Blake, N. K. & Pilgeram, A. L. (1988). Transformation of *Gaeumannomyces graminis* to benomyl resistance. *Current Genetics* **14**, 113-117.

Herrera-Estrella, A., Goldman, G. H. & Van Montagu, M. (1990). High efficiency transformation system for the biocontrol agents *Trichoderma* spp. *Molecular Microbiology* **4**, 839-843.

Horng, J. S., Linz, J. E. & Pestka, J. J. (1989). Cloning and characterization of the *trpC* gene from an aflatoxigenic strain of *Aspergillus parasiticus*. *Applied and Environmental Microbiology* **55**, 2561-2568.

Iimura, Y., Gomi, K., Uzu, H. & Hara, S. (1987). Transformation of *Aspergillus oryzae* through plasmid-mediated complementation of the methionine-auxotrophic mutation. *Agricultural and Biological Chemistry* **51**, 323-328.

John, M. A. & Peberdy, J. F. (1984). Transformation of *Aspergillus nidulans* using the *argB* gene. *Enzyme Microbial Technology* **6**, 386-389.

Kelly, J. M. & Hynes, M. J. (1985). Transformation of *Aspergillus niger* by the *amdS* gene of *Aspergillus nidulans*. *EMBO Journal* **4**, 475-479.

Kim, S. Y. & Marzluf, G. A. (1988). Transformation of *Neurospora crassa* with the *trp-1* gene and the effect of host strain upon the fate of the transforming DNA. *Current Genetics* **13**, 65-70.

Kinsey, J. A. & Rambosek, J. A. (1984). Transformation of *Neurospora crassa* with the cloned *am* (glutamate dehydrogenase) gene. *Molecular and Cellular Biology* **4**, 117-122.

Kolar, M., Punt, P. J., Van den Hondel, C. A. M. J. J. & Schwab, H. (1988). Transformation of *Penicillium chrysogenum* using dominant selection markers and expression of an *Escherichia coli lacZ* fusion gene. *Gene* 62, 127-134.

Kornegay, M. A. J. R., Pribnow, D. & Gold, M. H. (1989). Transformation by complementation of an adenine auxotroph of the lignin-degrading basidiomycete *Phanerochaete chrysosporium*. *Applied and Environmental Microbiology* **55**, 406-411.

Kos, A., Kuijvenhoven, J., Wernars, K., Bos, C. J., Van den Broek, H. W. J., Pouwels, P. H. & Van den Hondel, C. A. M. J. J. (1985). Isolation and characterization of the *Aspergillus niger trpC* gene. *Gene* **39**, 231-238.

Kronstad, J. W., Wang, J., Covert, S. F., Holden, D. W., McKnight, G. L. & Leong, S. A. (1989). Isolation of metabolic genes and demonstration of gene disruption in the phytopathogenic fungus *Ustilago maydis*. *Gene* **79**, 97-106.

Malardier, L., Daboussi, M. J., Julien, J., Roussel, F., Scazzocchio, C. & Brygoo, Y. (1989). Cloning of the nitrate reductase gene (*niaD*) of *Aspergillus nidulans* and its use for transformation of *Fusarium oxysporum*. *Gene* **78**, 147-156.

Mattern, I. E., Punt, P. J. & Van den Hondel, C. A. M. J. J. (1988). A vector of *Aspergillus* transformation conferring phleomycin resistance. *Fungal Genetics Newsletter* **35**, 25.

May, G. S., Gambino, J., Weatherbee, J. A. & Morris, N. R. (1985). Identification and functional analysis of β-tubulin genes by site-specific integrative transformation in *Aspergillus nidulans*. *Journal of Cell Biology* **100**, 712-718.

May, G. S., Waring, R. B., Osmani, S. A., Morris, N. R. & Denison, S. H. (1989). The coming of age of molecular biology in *Aspergillus nidulans*. In *Proceedings, EMBO Alko Workshop on Molecular Biology of Filamentous Fungi*, vol. 6, (ed. H. Nevailainen & M. Penttilä), pp. 11-20. Foundation of Biotechnical and Industrial Fermentation Research: Helsinki.

Mooibroek, H., Kuipers, A. G. J., Sietsma, J. H., Punt, P. J. & Wessels, J. G. H. (1990). Introduction of hygromycin B resistance into *Schizophyllum commune*: preferential methylation of donor DNA. *Molecular and General Genetics* **222**, 41-48.

Mullaney, E. J., Punt, P. J. & Van den Hondel, C. A. M. J. J. (1988). DNA mediated transformation of *Aspergillus ficuum*. *Applied and Microbial Biotechnology* **28**, 451-454.

Munoz-Rivas, A., Specht, C. A., Drummond, B. J., Froeliger, E., Novotny, C. P. & Ullrich, R. C. (1986). Transformation of the basidiomycete, *Schizophyllum commune*. *Molecular and General Genetics* **205**, 103-106.

Oakley, B. R., Rinehart, J. E., Mitchell, B. L., Oakley, C. E., Carmona, C., Gray, G. L & May, G. S. (1987a). Cloning, mapping and molecular analysis of the *pyrG* orotidine-5'-phosphate decarboxylase gene of *Aspergillus nidulans*. *Gene* **61**, 385-399.

Oakley, C. E., Weil, C. F., Kretz, P. L. & Oakley, B. R. (1987b). Cloning of the *riboB* locus of *Aspergillus nidulans*. *Gene* **53**, 293-298.

Oliver, R. P., Roberts, I. N., Harling, R., Kenyon, L., Punt, P. J., Dingemanse, M. A. & Van den Hondel, C. A. M. J. J. (1987). Transformation of *Fulvia fulva*, a fungal pathogen of tomato, to hygromycin B resistance. *Current Genetics* **12**, 231-233.

Orbach, M. J., Porro, E. B. & Yanofsky, C. (1986). Cloning and characterization of the gene for β-tubulin from a benomyl-resistant mutant of *Neurospora crassa* and its use as a dominant selectable marker. *Molecular and Cellular Biology* **6**, 2452-2461.

Osiewacz, H. D. & Weber, A. (1989). DNA mediated transformation of the filamentous fungus *Curvularia lunata* using a dominant selectable marker. *Applied and Microbial Biotechnology* **30**, 375-380.

Paietta, J. & Marzluf, G. A. (1985). Plasmid recovery from transformants and the isolation of chromosomal DNA segments improving plasmid replication in *Neurospora crassa*. *Current Genetics* **9**, 383-388.

Panaccione, D. G., McKiernan, M. & Hanau, R. M. (1988). *Colletotrichum graminicola* transformed with homologous and heterologous benomyl-resistance genes retains expected pathogenicity to corn. *Molecular plant-Microbe Interactions* **1**, 113-120.

Peberdy, J. F. (1989). Fungi without coats–protoplasts as tools for mycological research. *Mycological Research* **93**, 1-20.

Perrot, M., Barreau, C. & Bégueret, J. (1987). Nonintegrative transformation in the filamentous fungus *Podospora anserina*: stabilization of a linear vector by the chromosomal ends of *Tetrahymena thermophila*. *Molecular and Cellular Biology* **7**, 1725-1730.

Picknett, T. M., Saunders, G., Ford, P. & Holt, G. (1987). Development of a gene transfer system for *Penicillium chrysogenum*. *Current Genetics* **12**, 449-455.

Punt, P. J., Oliver, R. P., Dingemanse, M. A., Pouwels, P. H. & Van den Hondel, C. A. M. J. J. (1987). Transformation of *Aspergillus* based on the hygromycin B resistance marker from *Escherichia coli*. *Gene* **56**, 117-124.

Punt, P. J., Dingemanse, M. A., Kuyvenhoven, A., Soede, R. D. M., Pouwels, P. H. & Van den Hondel, C. A. M. J. J. (1990). Functional elements in the promoter region of the *Aspergillus nidulans gpdA* gene, encoding glyceraldehyde-3-phosphate dehydrogenase. *Gene* **93**, 101-109.

Queener, S. W., Ingolia, T. D., Skatrud, P. L., Chapman, J. L. & Kaster, K. R. (1985). A system for genetic transformation of *Cephalosporium acremonium*. In *Microbiology – 1985,* (ed. L. Leive), pp. 468-472.American Society of Microbiology: Washington, DC.

Radford, A., Pope, S., Scazi, A., Fraser, M. J. & Parish, J. H. (1981). Liposome-mediated genetic transformation of *Neurospora crassa*. *Molecular and General Genetics* **184**, 567-569.

Revuelta, J. L. & Jayaram, M. (1986). Transformation of *Phycomyces blakesleeanus* to G-418 resistance by an autonomously replicating plasmid. *Proceedings of the National Academy of Sciences, USA* **83**, 7344-7347.

Roberts, I. N., Oliver R. P., Punt P. J. & Van den Hondel, C. A. M. J. J. (1989). Expression of the *Escherichia coli* β-glucuronidase gene in industrial and phytopathogenic filamentous fungi. *Current Genetics* **15**, 177-180.

Samac, D. A. & Leong, S. A. (1989). Characterization of the termini of linear plasmids from *Nectria haematococcus* and their use in construction of an autonomously replicating transformation vector. *Current Genetics* **16**, 187-194.

Sánchez, E., Lozano, M., Rubio, V. & Penalva, M. A. (1987). Transformation in *Penicillium chrysogenum*. *Gene* **51**, 97-102.

Staben, C., Jensen, B., Singen, M., Pollock, J., Schechtman, M. & Kinsey, J. (1989). Use of a bacterial hygromycin B resistance gene as a dominant selectable marker in *Neurospora crassa* transformation. *Fungal Genetics Newsletter* **36**, 79-81.

Stinchcomb, D. T., Struhl, K. & Davis, R. W. (1979). Isolation and characterization of yeast chromosomal replicator. *Nature* **282**, 39-43.

Thomas, M. D. & Kenerly, C. M. (1989). Transformation of the mycoparasite *Gliocladium*. *Current Genetics* **15**, 415-420.

Tilburn, J., Scazzocchio, C., Taylor, G. G., Zabicky-Zissman, J. H., Lockington, R. A. & Davies, R. W. (1983). Transformation by integration in *Aspergillus nidulans*. *Gene* **26**, 205-221.

Timberlake, W. E. & Marshall, M. A. (1989). Genetic engineering of filamentous fungi. *Science* **244**, 1313- 1317.

Tsukuda, T., Carleton, S., Fotheringham, S. & Holloman, W. K. (1988). Isolation and characterization of an autonomously replicating sequence from *Ustilago maydis*. *Molecular and Cellular Biology* **8**, 3703-3709.

Turgeon, B. G., MacRae, W. D., Garber, R. C., Fink, G. R. & Yoder, O. C. (1986). A cloned tryptophan-synthesis gene from the Ascomycete *Cochliobolus heterostrophus* functions in *Escherichia coli*, yeast and *Aspergillus nidulans*. *Gene* **42**, 79-88.

Unkles, S. E., Campbell, E. I., Carrez, D., Grieve, C., Contreras, R., Fiers, W., Van den Hondel, C. A. M. J. J. & Kinghorn, J. R. (1989). Transformation of *Aspergillus niger* with the homologous nitrate reductase gene. *Gene* **78**, 157-166.

Unkles, S. E., Campbell, E. I., De Ruiter-Jacobs, Y. M. J. T., Broekhuijsen, M., Macro, J. A., Carrez, D., Contreras, R., Van den Hondel, C. A. M. J. J. & Kinghorn, J. R. (1989). The development of a homologous transformation system for *Aspergillus oryzae* based on the nitrate assimilation pathway: A convenient and general selection system for filamentous fungal transformation. *Molecular and General Genetics* **218**, 99-104.

Van Gorcom, R. F. M., Boschloo, J. G., Kuyvenhoven, A., Lange, J., Van Vark, A. J., Bos, C. J., Van Balken, J. A. M. & Van den Hondel, C. A. M. J. J. (1990). Isolation and molecular characterization of the benzoate-para-hydroxylase gene (*bphA*) of *A. niger* a member of a new gene-family of the cytochrome P450 superfamily. *Molecular and General Genetics* **223**, 192-197.

Van Gorcom, R. F. M., Pouwels, P. H., Goosen, T., Visser, J., Van den Broek, H. W. J., Hamer, J. E., Timberlake, W. E. & Van den Hondel, C. A. M. J. J. (1985). Expression of an *Escherichia coli* β-galactosidase fusion gene in *Aspergillus nidulans*. *Gene* **40**, 99-106.

Van Gorcom, R. F. M., Punt, P. J., Pouwels, P. H. & Van den Hondel, C. A. M. J. J. (1986). A system for the analysis of expression signals in *Aspergillus*. *Gene* **48**, 211-217.

Van Gorcom, R. F. M. & Van den Hondel, C. A. M. J. J. (1988). Expression analysis vectors for *Aspergillus niger*. *Nucleic Acid Research* **16**, 9052.

Van Hartingsveldt, W., Mattern, I. E., Van Zeijl, C. M. J., Pouwels, P. H. & Van den Hondel, C. A. M. J. J. (1987). Development of a homologous transformation system for *Aspergillus niger* based on the *pyrG* gene. *Molecular and General Genetics* **206**, 71-75.

Van Hartingsveldt, W., Van den Hondel, C. A. M. J. J., Veenstra, A. E. & Van den Berg, J. A. (1990). Gene replacement as a tool for the construction of *Aspergillus* strains. European Patent Application 89202106.4.

Van Heeswijck, R. (1986). Autonomous replication of plasmids in *Mucor* transformants. *Carlsberg Research Communications* **51**, 433-443.

Van Heeswijck, R. & Roncere, M. I. G. (1984). High frequency transformation of *Mucor* with recombinant plasmid DNA. *Carlsberg Research Communications* **49**, 691-702.

Vollmer, S. J. & Yanofsky, C. (1986). Efficient cloning of genes of *Neurospora crassa*. *Proceedings of the National Academy of Sciences, USA* **83**, 4869-4873.

Ward, M., Kodama, K. H. & Wilson, L. J. (1989). Transformation of *Aspergillus awamori* and *Aspergillus niger* by electroporation. *Experimental Mycology* **13**, 289-293.

Ward, M., Wilkinson, B. & Turner, G. (1986). Transformation of *Aspergillus nidulans* with a cloned oligomycin resistant ATP synthase subunit 9 gene. *Molecular and General Genetics* **202**, 265-270.

Ward, M., Wilson, L. J., Carmona, C. L. & Turner, G. (1988). The *oliC3* gene of *Aspergillus niger*: isolation, sequence and use as a selectable marker for transformation. *Current Genetics* **14**, 37-42.

Wernars, N., Goosen, T., Wennekes, L. M. J., Visser, J., Bos, C. J., Van den Broek, H. W. J., Van Gorcom, R. F. M., Van den Hondel, C. A. M. J. J. & Pouwels, P. H. (1985). Gene amplification in *Aspergillus nidulans* by transformation with vectors containing the *amdS* gene. *Current Genetics* **9**, 361-368.

Whitehead, M. P., Unkles, S. E., Ramsden, M., Campbell, E. I., Gurr, S. J., Spence, D., Van den Hondel, C. A. M. J. J., Contreras, R. & Kinghorn, J. R. (1989). Transformation of a nitrate reductase deficient mutant of *Penicillium chrysogenum* with the corresponding *Aspergillus niger* and *A. nidulans niaD* genes. *Molecular and General Genetics* **216**, 408-411.

Wnendt, F., Jacobs, M. & Stahl, U. (1990). Transformation of *Aspergillus giganteus* to hygromycin B resistance. *Current Genetics* **17**, 21-24.

Wöstemeyer, J., Burmester, A. & Weigel, C. (1987). Neomycin resistance as a dominantly selectable marker for transformation of the zygomycete *Absidia glauca*. *Current Genetics* **12**, 625-627.

Yelton, M. M., Hamer, J. E. & Timberlake, W. E. (1984). Transformation of *Aspergillus nidulans* by using a *trpC* plasmid. *Proceedings of the National Academy of Sciences, USA* **81**, 1470-1474.

Chapter 2

Strategies for cloning genes from filamentous fungi

Geoffrey Turner

Since the first chromosomal genes were isolated from filamentous fungi (e.g. Vapnek *et al.*, 1977), there has been a rapid growth in the number of genes reported isolated from a wide variety of fungi. *Aspergillus nidulans* and *Neurospora crassa* represent by far the main organisms in this respect, partly because they were the major species with which the earlier classical genetic studies were carried out. As the list of cloned genes has grown over the last 13 years, it has extended to a wide range of species. Lists of cloned genes and a bibliography are printed in the Fungal Genetics Newsletter, obtainable from the Fungal Genetics Stock Center (Department of Microbiology, University of Kansas Medical Center, Kansas City, Kansas 66103, U.S.A.). This information is updated annually and includes the names and addresses of cloners. For basic methodology in genetic manipulation techniques and vector design, the reader should refer to Old & Primrose (1990), and for practical details to Sambrook, Fritsch & Maniatis, 1989.

When fungal gene isolation is contemplated, two points should be remembered. First, for structural genes, sequence similarity often permits a gene from one fungus to be used as a probe to detect sequences from another species. Second, data from many studies now show that a gene from any Euascomycete is usually transcribed and often expressed, without any modification of the gene promoter, in other Euascomycetes (or related Deuteromycete). This phenomenon is often described as 'heterologous expression', and probably reflects a basic similarity in both promoter sequences and intron splice sites within this class. As yet, there are insufficient data to reach any firm conclusions about gene transfer between different classes, such as Ascomycetes and Basidiomycetes.

The mitochondrial genomes of a number of filamentous fungi have also been well characterized, and in some cases completely sequenced. This work is fully reviewed elsewhere (Brown, 1987).

The major methods used to date for cloning fungal genes are outlined in Table 2.1. By 1989, approximately 120 genes had been cloned from *A. nidulans* and *N. crassa* alone, so no attempt has been made to give a complete bibliography.

Table 2.1. A summary of cloning strategies

Heterologous expression in *Escherichia coli* or yeast
Expression in filamentous fungi
Differential hybridization with cDNA
cDNA expression and protein detection
Chromosome aberrations
Oligonucleotide probes designed from protein sequence
Heterologous probing
Chromosome walking

Construction of gene libraries

Cosmid, plasmid and bacteriophage vectors have been used, as appropriate, to construct libraries from total fungal DNA or cDNA (reverse transcribed from fungal mRNA). Although many standard vectors are used, certain cosmid and plasmid vectors have been specially constructed to carry suitable markers for selection of fungal transformants (Table 2.2). These are especially helpful if cloning by complementation of fungal mutations is intended. Bacteriophage lambda vectors such as EMBL4 are useful for heterologous probing, and lambda gt11 for cDNA cloning and expression.

More recently, progress has been made towards the development of ordered gene libraries, using cosmid vectors. Each cosmid clone has a specific number. As information becomes available from many laboratories on the identity of clones carrying particular genes, on the chromosomal origin of individual clones, and on the overlap between different clones, the library can be ordered, and a physical picture of the whole genome is built up. This approach has already been started for *N. crassa* (Ballario *et al.*, 1989), and is being extended to *A. nidulans* (W. E. Timberlake, personal communication). These cosmid banks and information about them can be obtained from the Fungal Genetics Stock Center. The technique of chromosome separation by pulsed field gel electrophoresis (Orbach *et al.*, 1988) and possibilities for long range mapping using rare cutting restriction endonuclease enzymes, will contribute to obtaining complete physical maps for many nuclear genomes. The combination of genetic and physical data will then greatly facilitate the isolation of genes.

Table 2.2. Plasmid (P) and cosmid (C) cloning vectors for filamentous fungi

Vector	type	marker gene	host	gene isolation method	reference
pILJ16	P	*argB*	*Aspergillus nidulans*	rescue	1
pDJB3	P	*pyr-4*	*A. nidulans*	rescue	2
pKBY2	C	*trpC*	*A. nidulans*	packaging	3
			Nectria haematococca	packaging	3
pBR329	P	none	*A. nidulans*	rescue	4
pRAL1	P	*qa-2*	*Neurospora crassa*	sib-selection	5
pSV50	C	*tub-2*	*N. crassa*	sib-selection	6

References: 1, Johnstone *et al.*, 1985; 2, Ballance & Turner, 1986; 3, Yelton *et al.*, 1985, Weltring *et al.*, 1988; 4, Oakley *et al.*, 1987; 5, Akins & Lambowitz, 1985; 6, Vollmer & Yanofsky, 1986.

Gene isolation

Heterologous expression

Escherichia coli

Transformation and complementation of *E. coli* auxotrophic mutants with fungal DNA libraries constructed in plasmid vectors was used to isolate some of the first cloned fungal genes (Vapnek *et al.*, 1977; Kinghorn & Hawkins, 1982). It depended to a very high degree on luck, since fungal promoters are not recognised in bacteria, leading to poor or negligible transcription and translation. Introns, if present, cannot be excised from the primary transcript in *E. coli*, leading to truncated proteins. Nevertheless, for certain intronless genes, weak expression is sufficient to complement a nutritional lesion in the bacterium. Genes of several fungi complementing *trpC* mutants of *E. coli* have been cloned by this approach, and usually encode a multifunctional polypeptide responsible for 3 steps in tryptophan biosynthesis (e.g. Schechtman & Yanofsky, 1983; Yelton *et al.*, 1983; Choi *et al.*, 1988). Similarly, the *pyr-4* gene of *N. crassa* was cloned by complementation of a *pyrF* mutant of *E. coli* (Buxton & Radford, 1983).

Saccharomyces cerevisiae

In the early 1980s, before the advent of filamentous fungal cloning systems, a large amount of unrewarded effort was expended on attempting to clone fungal genes by expression in *E. coli* and yeast (e.g. Ullrich *et al.*, 1985). Promoter and intron recognition problems have, rather disappointingly, made *S. cerevisiae* little better than *E. coli* as a cloning host for the isolation of genes from filamentous fungi. The *argB* gene of *A. nidulans* is one of the few genes which has been cloned by expression in yeast (Berse *et al.*, 1983). The *pyr-4* gene of *N. crassa*, isolated by expression in *E. coli*, also complements the *ura3* mutation of yeast (Ballance, Buxton & Turner, 1983).

Expression in filamentous fungi

The problems associated with attempts to clone genes in bacteria and yeast soon led to the development and improvement of fungal transformation systems, fully reviewed elsewhere (Fincham, 1989; Chapter 1). Table 2.2 lists some of the vectors which are now used for gene isolation by transformation of filamentous fungi. Cloning by this approach depends on the availability of a relatively efficient transformation system, and/or large inserts in the vector, achieved by use of a cosmid-based vector. For a simple plasmid vector, which might have an average insert size of about 8 kb, 18000 *E. coli* clones are required to be 99% certain of finding any sequence in a library of *A. nidulans*, where the genome size is 31 megabases (Brody & Carbon, 1988). For a cosmid vector, with average insert size of about 35 kb, only 4000 clones are needed for 99% certainty. To convert a plasmid into a cosmid simply requires addition of the lambda *cos* sites, required for packaging in phage particles prior to transfection. Corresponding numbers of fungal transformants, using library DNA, are required to isolate a given gene by complementation of a mutant.

These requirements have now been fulfilled for *A. nidulans* and *N. crassa* and more recently for *Podospora anserina* (Begueret *et al.*, 1988) and *Coprinus cinereus* (Casselton *et al.*, 1989). For many other fungi, especially most plant pathogens, transformation frequency remains at 1 to 20 per μg of DNA, and even the cosmid approach has made little headway to date. Here, gene cloning has depended upon heterologous probing or cDNA approaches. Complementation of a heterologous filamentous fungal host as a method for isolating a gene has been limited, the best example being the isolation of pisatin demethylase from the plant pathogen *Nectria haematococca* by expression in *A. nidulans* (Weltring *et al.*, 1988).

Once a fungal transformant has been obtained, in which it appears that an insert sequence is complementing the mutation of interest, it is necessary to recover this sequence. In *A. nidulans*, rescue of the transforming sequence from total transformant DNA is the usual method. As in most filamentous fungi, the transforming sequence is integrated into the genome of the host. Since the complementing insert is adjacent to the bacterial vector sequences (usually the ampicillin resistance marker and plasmid origin of replication), partial digestion of total transformant

DNA, religation, and transformation of *E. coli* to ampicillin resistance ('marker rescue') results in recovery of the desired sequence. Common complications include plasmid rearrangements, and recovery of plasmids not containing the desired sequence. More than one plasmid type can integrate in the same nucleus during transformation, i.e. cotransformation is frequent. It is therefore necessary to test a number of rescued plasmids by retransformation of the fungal mutant (Johnstone *et al.*, 1985; Ballance & Turner, 1986).

When cosmid libraries are used for gene isolation by this method, undigested DNA from the fungal transformant can be added to a lambda packaging mixture, and *E. coli* transfected. This results in the recovery of cosmid-sized molecules, some of which retransform the fungal mutant (Yelton, Timberlake & Van den Hondel, 1985). It is probable that a small number of free cosmid molecules are excised from the genome by rare recombination events between adjacent duplicated sequences, and these molecules are recovered by the packaging procedure. It is known that such recombination events do occur during vegetative growth (Ward, Wilkinson & Turner, 1986). Some of these rescued cosmids contain the wild type allele, and some the mutant allele of the desired gene. They cannot usually be distinguished by restriction endonuclease digestion, and must be tested by transformation. In place of marker rescue, *Neurospora* workers have used a technique known as 'sib-selection' (Akins & Lambowitz, 1985; Vollmer & Yanofsky, 1986), where individual cosmid or plasmid clones are pooled in a number of groups. DNA prepared from each group is used to transform the desired mutant, looking for complementation. Once a group is identified which contains the required sequence, it is subdivided, and the process repeated. Eventually, a single cosmid clone responsible for the positive transformation is identified. This technique avoids the need to rescue sequences from the fungus. Since the standard technique for obtaining homokaryotic strains of *N. crassa* is *via* ascospores, and sexual crosses can inactivate duplicated gene sequences arising from transformation (the RIP phenomenon, Selker & Garrett, 1988), sib-selection offers some advantages for this organism. Conversely, marker rescue has been the preferred method to date for *A. nidulans*, where uninucleate conidiospores provide an easy method for isolating homokaryotic transformants.

When a cosmid clone has been identified which seems to contain the required gene, it is necessary to locate the gene, which might be just a few kilobases, within the large cosmid insert (35 kb or more). This can be achieved by cotransforming the mutant recipient with a simple selectable vector and purified restriction endonuclease fragments derived from the cosmid, and looking for fragments which complement the mutation (Timberlake *et al.*, 1985).

Differential hybridization with cDNA

This approach has been used to isolate genes which are expressed under one set of circumstances, but not another. Messenger RNA is prepared from mycelium grown under each condition, and reverse transcribed to produce a labelled probe of cDNA. In one approach, duplicate filters, lifted from gene library (cDNA or genomic) plaques or colonies, are then challenged with each probe, and comparisons made. Clones carrying sequences which are transcribed under both conditions are detected on both filters, while differentially transcribed sequences are found on only one filter. Growth conditions which are appropriate for this approach include acetate/sucrose for acetate-inducible genes (Thomas, Connerton & Fincham, 1988; Sandeman & Hynes, 1989), exponential phase/idiophase for lignin peroxidase (Tien & Tu, 1987; Brown *et al*., 1988), and vegetative/sporulating mycelium (Timberlake & Barnard, 1981). Once clones of differentially-expressed genes are obtained, the problem of gene identification still remains, and various approaches are possible (see below).

Complementary-DNA expression and protein detection

This approach, which has been used extensively with higher organisms and plants, has been used to a limited extent with fungi. Although more than one approach is possible, it requires preparation of antibody to the purified protein of interest. Either cDNA is cloned into an expression vector such as lambda gt11, for expression in *E. coli*, or pooled DNA from individual cDNA clones is used to select homologous RNA, which is expressed in an *in vitro* translation system (hybrid release translation). In both cases, the protein, if expressed, is then detected immunologically, permitting recognition of the required clones.

Hybrid release translation was used to isolate cDNA for the gene encoding subunit 9 of the mitochondrial ATP synthetase of *N. crassa* (Viebrock, Perz & Sebald, 1982); an alkaline protease from *Aspergillus oryzae* was cloned by a similar approach (Tatsumi *et al*., 1989).

Chromosome aberrations

In *A. nidulans*, where collections of mapped translocation strains are available, methods have been devised to take advantage of chromosome aberrations which result in mutations in the gene of interest. A rather laborious technique was used successfully before transformation cloning became available (Green & Scazzocchio, 1985). The method involved screening for restriction fragment differences between wild type and a translocation strain, where the translocation break point was within the gene of interest. In effect, this meant probing the wild type and translocation mutant with every clone in a lambda library until the break point was detected as a difference in the restriction pattern following hybridization with the probe. Since repeating this for each gene would have been very tedious, a strain was constructed which contained a number of mutations and translocations. When ordered cosmid libraries, with chromosome-specific clones, become more easily available, this approach may find further applications.

Loss of sequence, in duplication deficient progeny from a cross between a strain carrying a mutation resulting from a pericentric inversion and an aberration free strain, was used to clone the *areA* regulatory gene of *A. nidulans* (Caddick *et al.*, 1986).

Oligonucleotide probes designed from protein sequence

If sufficient purified protein can be obtained, it is possible to determine its amino acid sequence, and to use this sequence information to synthesize an oligonucleotide probe. An early example of this technique applied to a fungus was the isolation of the *am* gene of *N. crassa*, encoding NADP-dependent glutamate dehydrogenase (Kinnaird *et al.*, 1982). A partial sequence of the purified protein was obtained, and oligonucleotide mixtures designed and synthesized from the least ambiguous regions. As the technology for protein sequencing and oligonucleotide synthesis has improved and become more widely available, this approach has been used occasionally for fungal gene isolation, most notably for the cloning of some of the genes involved in antibiotic biosynthesis from *Acremonium chrysogenum* (Samson *et al.*, 1985) and *Penicillium chrysogenum* (Barredo *et al.*, 1989).

Heterologous probing

Nucleotide sequences of equivalent genes are often similar, so that a gene from one organism can be used to identify the equivalent gene from a second organism by hybridization. As might be expected, this similarity is roughly related to the phylogenetic relationship of the organisms concerned. Thus the phosphoglycerate kinase gene of *Saccharomyces cerevisiae* has been used as a probe to isolate the equivalent *A. nidulans* gene (Clements & Roberts, 1986), but it is less likely that the equivalent bacterial or mammalian gene could be employed in the same way to isolate the fungal gene. Some genes are conserved much more than others over a wide range of species, ribosomal RNA, histone and tubulin genes being good examples. For instance, *Drosophila* and chicken genes have been used as heterologous probes to identify histone and tubulin genes, respectively, in *A. nidulans* (May & Morris, 1987; May *et al.*, 1987). The *oliC* gene of *A. nidulans* encodes subunit 9 of the mitochondrial ATP synthetase, a highly conserved complex throughout nature, and was isolated using the *N. crassa* cDNA clone as a probe (Ward *et al.*, 1986). The coding sequences of the respective genes show 75% nucleotide sequence identity.

Subsequently, the *A. nidulans* gene was used to isolate equivalent genes from *A. niger* (Ward *et al.*, 1988) and *Penicillium chrysogenum* (Bull, Smith & Turner, 1988). In contrast, the *pyr-4* gene of *N. crassa* has only 49% identity with the equivalent *A. nidulans pyrG* gene, and it was not possible to find the latter by probing with the former. Thus success with this method is often a matter of trial and error. For instance, in attempting to isolate the acetyl CoA synthetase gene from *Ustilago maydis* (a hemibasidiomycete) for use as a transformation marker, when the equivalent *A. nidulans* gene failed as a probe, the *N. crassa* gene was used successfully (Hargreaves & Turner, 1989). More surprisingly, it has been possible to identify and map genes encoding synthesis of β-lactam antibiotics in *Penicillium chrysogenum* and *A. nidulans* using probes from a similar pathway in a Gram negative *Flavobacterium* species (Smith *et al.*, 1990b). The high nucleotide sequence similarity of these antibiotic biosynthesis genes between such widely divergent organisms is unusual, and has led to a hypothesis of horizontal gene transfer long after the fungi and bacteria diverged (Ramon *et al.*, 1987; Weigel *et al.*, 1988; Landan *et al.*, 1990; see also Chapter 9). Since isolation of genes by heterologous probing is probably the simplest method of all, this approach has been used widely between yeast and Euascomycetes (Clements & Roberts, 1986; Punt *et al.*, 1988), and within and between other fungal classes (Gurr *et al.*, 1986; Mellon *et al.*, 1987; Hartingsveld *et al.*, 1987; Sims *et al.*, 1988; Cooley & Caten, 1989). The method has been partly responsible for the rapid extension of molecular information/technology from one fungal species to another.

Chromosome walking

For fungal species where a genetic map is available, it is sometimes possible, having isolated one gene or sequence, to 'walk' along the chromosome to an adjacent gene. This method was exploited in the isolation of the *qa* cluster of *N. crassa*, once one of the genes had been isolated by complementation of an *E. coli* mutant (Lamb *et al.*, 1990).

The term 'walking' refers to the sequential isolation of overlapping clones, using the first clone to probe the gene library in order to isolate further clones carrying parts of the same insert together with adjacent sequences. This 'extends' the available sequences in both directions from the initial isolate. Since there is no strict relationship between genetic distance in centimorgans and physical distance in kilobases of DNA, it is not possible to predict exactly how far apart any two closely linked genes are. However, some data are now available. The *acu*D$^+$ gene of *A. nidulans*, encoding isocitrate lyase, was cloned by complementation of a host mutation (Ballace & Turner, 1986), and was only about 1 cM from *fac*A, encoding acetyl CoA synthetase (Armitt, McCullough & Roberts,

1976). Data from total physical genome size and total size in cM suggested that the distance between them might be only a few kilobases. Walking using lambda and cosmid clones eventually demonstrated that they are separated by about 30 kb (J. Brown & G. Turner, unpublished). The unexpectedly low frequency of recombination in this region does not hold true for other loci (e.g. Hull *et al.*, 1989). These data reflect the uneven pattern of recombination across the genome of *A. nidulans*. One of the mating type factors (A3) of the basidiomycete *Coprinus cinereus* was recently cloned by walking from the adjacent *paba-1* locus (Casselton *et al.*, 1989); although only 0·5 cM apart, the physical distance was 47 kb.

Walking is facilitated if the library is constructed in a cosmid vector, since larger inserts lead to faster walking. Since labelled probes from the ends of the insert are best for further extending the cloned region, some method of easily labelling the ends of the insert is desirable, avoiding the laborious task of having to map the whole clone and isolate the insert ends. Examples of such vectors are Lorist 2 (Cross & Little, 1986) and the pWE series (Wahl *et al.*, 1987). These carry bacteriophage transcriptional promoters within the vector sequences, close to the insert, and in such an orientation that synthesis of labelled DNA is directed into the adjacent insert DNA.

Identification of cloned genes

When a putative gene sequence has been isolated by one of the methods listed above, its identity has to be confirmed. If the host fungus is well characterized genetically, as in the case of *A. nidulans*, the identity of genes cloned by transformation and complementation (or suppression) of a known mutation is generally confirmed by correlating molecular and genetic mapping data. A cloned sequence is inserted at the homologous host sequence by transformation, the site of integration being checked by hybridization analysis. The integration site is then mapped genetically by a sexual cross, to confirm that the sequence corresponds to the known genetic locus (e.g. Yelton *et al.*, 1985; Ballance & Turner, 1986).

Sequences isolated by differential hybridization may represent a number of different genes, all expressed under the chosen conditions, and the exact identity of each gene may therefore be unknown. Genes isolated by probing with a gene from another organism must also be checked for their true identity. How can the genes be identified? Methods include transformation of appropriate mutants to look for complementation, expression of the cloned sequence in a suitable host and examination of the nature of the product (e.g. enzyme assay), or disruption of the sequence in the host fungus by transformation, and looking for phenotypic effects of the null mutation (Fincham, 1989; see Chapter 1).

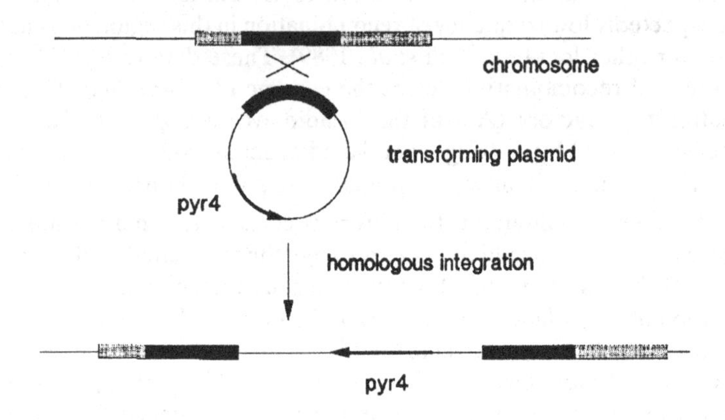

Fig. 2.1. Gene disruption by transformation. A restriction endonuclease fragment internal to the gene is cloned into a transformation vector. Transformation of the wild type strain by homologous recombination results in the insertion of the vector sequences within the gene of interest, thus inactivating it.

For example, sequences expressed specifically during acetate induction in *A. nidulans* or *N. crassa* could represent genes for acetate uptake, acetyl CoA synthetase, or glyoxylate cycle enzymes. The identity of sequences isolated by this approach was established by transformation and complementation of appropriate mutants, or detection of a translocation break point, resulting in a mutation in the gene of interest, by hybridization of the isolated sequence with DNA extracted from the mutant (Thomas *et al.*, 1988; Sandeman & Hynes, 1989). Expression of a gene or group of genes in a related organism which naturally lacks the function of interest can also be used to confirm function. For example, a cluster of genes believed to encode the entire secondary biosynthetic pathway for penicillin, isolated from *P. chrysogenum*, was introduced into *N. crassa* and *A. niger*. These species do not produce penicillin, but the transformed strains do (Smith *et al.*, 1990a). Gene disruption was also used to show that a sequence suspected of encoding the first step in penicillin biosynthesis was essential for penicillin production (Smith *et al.*, 1990b). Gene disruption was achieved as shown in Fig. 2.1. Since non-homologous (ectopic) integration is also possible, digested DNA of transformants must be analyzed by Southern blotting to check the nature of the integration event.

Sequences isolated using oligonucleotide probes designed from determined amino acid sequence have often turned out to be the desired gene (Kinnaird *et al.*, 1982; Samson *et al.*, 1985), but it is also common to obtain false signals. Because of redundancy in the genetic code, it is only possible to design an oligonucleotide which approximates to the correct nucleotide sequence of the gene. Once a hybridizing sequence has been isolated, it must be sequenced to check that it does encode the determined amino acid sequence.

Conclusion

The wide range of methods now available for the isolation of fungal genes is leading to a rapid growth in published sequence data from many fungal species, and such information should be invaluable to fungal taxonomists. In addition to publication in journals, sequences are submitted to DNA sequence databases in Europe and the U.S.A., known as EMBL and GenBank, and any mycologist with access to a computer network can now rapidly recover sequences and carry out sequence comparisons with the appropriate software.

Cloning and sequencing of genes is only the first step in learning about their function and expression, and transformation techniques for fungi (Chapter 1) now provide us with the means of asking many important questions about all types of fungal genes. All areas of mycology should benefit greatly from these developments.

References

Akins, R. A. & Lambowitz, A. M. (1985). General method for cloning *Neurospora crassa* nuclear genes by complementation of mutants. *Molecular and Cellular Biology* **5**, 2272-2278.

Armitt, S., McCullough, W. & Roberts, C. F. (1976). Analysis of acetate non-utilizing (*acu*) mutants in *Aspergillus nidulans*. *Journal of General Microbiology* **92**, 263-282.

Ballance, D. J., Buxton, F. P. & Turner, G. (1983). Transformation of *Aspergillus nidulans* by the orotidine-5'-phosphate decarboxylase gene of *Neurospora crassa*. *Biochemical and Biophysical Research Communications* **112**, 284-289.

Ballance, D. J. & Turner, G. (1986). Gene cloning in *Aspergillus nidulans*: isolation of the isocitrate lyase gene (*acu*D). *Molecular and General Genetics* **202**, 271-275.

Ballario, P., Morelli, G., Sporeno, E. & Macino, G. (1989). Cosmids from the Vollmer-Yanofsky library identified with a chromosome VII probe. *Fungal Genetics Newsletter* **36**, 38-39.

Barredo, J. L., Van Solingen, P., Diéz, B., Alvarez, E., Cantoral, J. M., Kattevilder, A., Smaal, E. B., Groenen, M. A. M., Veenstra, A. E. & Martín, J. F. (1989). Cloning and characterization of the acyl-coenzyme a: 6-aminopenicillanic acid-acyltransferase gene of *Penicillium chrysogenum*. *Genetics* **83**, 291-300.

Begueret, J., Turcq, B., Razanamparany, V., Perriere, M., Denayrolles, M., Berges, T., Perrot, M., Javerzat, J. -P. & Barreau, C. (1989). Development and use of vectors for *Podospora anserina*. In *Molecular Biology of Filamentous Fungi*, (ed. H. Nevalainen & M. Penttilä), pp. 41-49. Foundation for Biotechnical and Biochemical Research: Helsinki.

Berse, B., Dmochowska, A., Skrzypek, M., Weglenski, P., Bates, M. A. & Weiss, R. L. (1983). Cloning and characterization of the ornithine carbamoyl-transferase gene from *Aspergillus nidulans*. *Genetics* **25**, 109-117.

Brody, H. & Carbon, J. (1989). Electrophoretic karyotype of *Aspergillus nidulans*. *Proceedings of the National Academy of Sciences, USA* **86**, 6260- 6263.

Brown, A., Sims, P., Raeder, U. & Broda, P. (1988). Multiple ligninase-related genes from *Phanerochaete chrysosporium*. *Genetics* **73**, 77-85.

Brown, T. A. (1987). The mitochondrial genomes of filamentous fungi. In *Gene Structure and Function in Eukaryotic Microbes*, (ed. J. R. Kinghorn), pp. 141-162. IRL Press: Oxford and Washington.

Bull, J. H., Smith, D. J. & Turner, G. (1988). Transformation of *Penicillium chrysogenum* with a dominant selectable marker. *Current Genetics* **13**, 377-382.

Buxton, F. P. & Radford, A. (1983). Cloning of the structural gene for orotidine-5'-phosphate carboxylase of *Neurospora crassa* by expression in *Escherichia coli*. *Molecular and General Genetics* **190**, 403-405.

Caddick, M. X., Arst, H. N. Jr., Taylor, L. H., Johnson, R. I. & Brownlee, A. G. (1986). Cloning of the regulatory gene *areA* mediating nitrogen metabolite repression in *Aspergillus nidulans*. *EMBO Journal* **5**, 1087-1090.

Casselton, L. A., Mutasa, E. S., Tymon, A., Mellon, F. M., Little, P. F. R., Taylor, S., Bernhagen, J. & Stratman, R. (1989). The molecular analysis of Basidiomycete mating type genes. In *Molecular Biology of Filamentous Fungi*, (ed. H. Nevalainen & M. Penttilä), pp. 139-148. Foundation for Biotechnical and Biochemical Research: Helsinki.

Choi, H. T., Revuelta, J. L., Sadhu, C. & Jayaram, M. (1988). Structural organization of the *TRP1* gene of *Phycomyces blakesleeanus*: Implications for evolutionary gene fusion in fungi. *Genetics* **71**, 85-95.

Clements, J. M. & Roberts, C. F. (1986). Transcription and processing signals in the 3-phosphoglycerate kinase (PGK) gene from *Aspergillus nidulans*. *Genetics* **44**, 97-105.

Cooley, R. N. & Caten, C. E. (1989). Cloning and characterization of the β-tubulin gene and determination of benomyl resistance in *Septoria nodorum*. In *Molecular Biology of Filamentous Fungi*, (ed. H. Nevalainen & M. Penttilä), pp. 207-216. Foundation for Biotechnical and Biochemical Research: Helsinki.

Cross, S. H. & Little, P. F. R. (1986). A cosmid vector for systematic chromosome walking. *Genetics* **49**, 9-22.

Fincham, J. R. S. (1989). Transformation in fungi. *Microbiological Reviews* **53**, 148-170.

Green, P. M. & Scazzocchio, C. (1985). A cloning strategy in filamentous fungi. In *Gene Manipulation in Fungi*, (ed. J. W. Bennett & L. A. Lasure), pp. 345-353. Academic Press: Orlando & London.

Gurr, S. J., Hawkins, A. R., Drainas, C. & Kinghorn, J. R. (1986). Isolation and identification of the *Aspergillus nidulans gdhA* gene encoding NADP-linked glutamate dehydrogenase. *Current Genetics* **10**, 761-766.

Hargreaves, J. A. & Turner, G. (1989). Isolation of the acetyl-CoA synthase gene from the corn smut pathogen, *Ustilago maydis*. *Journal of General Microbiology* **135**, 2675-2678.

Hull, E. P., Green, P. M., Arst, H. N. Jr. & Scazzocchio, C. (1989). Cloning and physical characterization of the L-proline catabolism gene cluster of *Aspergillus nidulans*. *Molecular Microbiology* **3**, 553-559.

Johnstone, I. L., Hughes, S. G. & Clutterbuck, A. J. (1985). Cloning an *Aspergillus nidulans* developmental gene by transformation. *EMBO Journal* **4**, 1307-1311.

Kinghorn, J. R. & Hawkins, A. R. (1982). Cloning and expression in *Escherichia coli* K12 of the biosynthetic dehydroquinase function of the *arom* cluster gene from the eukaryote *Aspergillus nidulans*. *Molecular and General Genetics* **186**, 145-152.

Kinnaird, J. H., Keighren, M. A., Eaton, M. & Fincham, J. R. S. (1982). Cloning of the *am* (glutamate dehydrogenase) gene of *Neurospora crassa* through the use of a synthetic DNA probe. *Genetics* **20**, 387-396.

Lamb, H. K., Hawkins, A. R., Smith, M., Harvey, I. J., Brown, J., Turner, G., & Roberts, C. F. (1990). Spatial and biological characterisation of the complete quinic acid utilisation gene cluster in *Aspergillus nidulans*. *Molecular and General Genetics* **223**, 17-23.

Landan, G., Cohen, G., Aharonowitz, Y., Shuali, Y., Graur, D. & Shiffman, D. (1990). Evolution of isopenicillin N synthase genes may have involved horizontal gene transfer. *Molecular Biology and Evolution* **7**, 399-406.

May G. S. & Morris, N. R. (1987). The unique histone H2A gene of *Aspergillus nidulans* contains three introns. *Genetics* **58**, 59-66.

May, G. S., Tsang, M. L. -S., Smith, H., Fidel, S. & Morris, N. R. (1987). *Aspergillus nidulans* β-tubulin genes are unusually divergent. *Genetics* **55**, 231-243.

Mellon, F. M., Little, P. F. R. & Casselton, L. A. (1987). Gene cloning and transformation in the basidiomycete fungus *Coprinus cinereus*: isolation and expression of the isocitrate lyase gene (*acu-7*). *Molecular and General Genetics* **210**, 352-357.

Oakley, C. E., Weil, C. F., Kretz, P. L. & Oakley, B. R. (1987). Cloning of the *ribo*B locus of *Aspergillus nidulans*. *Genetics* **53**, 293-298.

Old, R. W. & Primrose, S. B. (1989). *Principles of Gene Manipulation*, 4th edn. Blackwell Scientific Publications: Oxford.

Orbach, M. J., Vollrath, D., Davis, R. W. & Yanofsky, C. (1988). An electrophoretic karyotype of *Neurospora crassa*. *Molecular Cell Biology* **8**, 1469-1473.

Punt, P. J., Dingemanse, M. A., Jacobs-Meijsing, B. J. M., Pouwels, P. H. & Van den Hondel, C. A. M. J. J. (1988). Isolation and characterization of the glyceraldehyde-3-phosphate dehydrogenase gene of *Aspergillus nidulans*. *Genetics* **69**, 49-57.

Ramón, D., Carramolino, L., Patino, C., Sanchez, F. & Peñalva, M. A. (1987). Cloning and characterization of the isopenicillin N synthetase gene mediating formation of the β-lactam ring in *Aspergillus nidulans*. *Genetics* **57**, 171-181.

Sambrook, J., Fritsch, E. F. & Maniatis, T. (1989). *Molecular Cloning - A Laboratory Manual*. 2nd edn. Cold Spring Harbor Laboratory Press: Cold Spring Harbor, U.S.A.

Samson, S. M., Belagaje, R., Blankenship, D. T., Chapman, J. L., Perry, D., Skatrud, P. L., Van Frank, R. M., Abraham, E. P., Baldwin, J. E. Queener, S. W. & Ingolia, T. D. (1985). Isolation, sequence determination and expression in *Escherichia coli* of the isopenicillin N synthetase gene from *Cephalosporium acremonium*. *Nature* **318**, 191-194.

Sandeman, R. A. & Hynes, M. J. (1989). Isolation of the *fac*A (acetyl-Coenzyme A synthetase) and *acu*E (malate synthase) genes of *Aspergillus nidulans*. *Molecular and General Genetics* **218**, 87-92.

Schechtman, M. G. & Yanofsky, C. (1983). Structure of the trifunctional *trp*-1 gene from *Neurospora crassa* and its aberrant expression in *E. coli*. *Journal of Molecular and Applied Genetics* **2**, 83-99.

Selker, E. U. & Garrett, P. W. (1988). DNA sequence duplications trigger gene inactivation in *Neurospora crassa*. *Proceedings of the National Academy of Sciences, USA* **85**, 6870-6874.

Sims, P., James, C. & Broda, P. (1988). The identification, molecular cloning and characterisation of a gene from *Phanerochaete chrysosporium* that shows strong homology to the exo-cellobiohydrolase I gene from *Trichoderma reesei*. *Genetics* **74**, 411-422.

Smith, D. J., Burnham, M. K. R., Edwards, J., Earl, A. J. & Turner, G. (1990a). Cloning and heterologous expression of the penicillin biosynthetic gene cluster from *Penicillium chrysogenum*. *Bio/Technology* **8**, 39-41.

Smith, D. J., Burnham, M. K. R., Bull, J. H., Hodgson, J. E., Ward, J. M., Browne, P., Brown, J., Barton, B., Earl, A. J. & Turner, G. (1990b). β-Lactam antibiotic biosynthetic genes have been conserved in clusters in prokaryotes and eukaryotes. *EMBO Journal* **9**, 741-747.

Tatsumi, H., Ogawa, Y., Murakami, S., Ishida, Y., Murakami, K., Masaki, A., Kawabe, H., Arimura, H., Nakano, E. & Motai, H. (1989). A full length cDNA clone for the alkaline protease from *Aspergillus oryzae*: structural analysis and expression in *Saccharomyces cerevisiae*. *Molecular and General Genetics* **219**, 33-38.

Thomas, G. H., Connerton, I. F. & Fincham, J. R. S. (1988). Molecular cloning and identification and transcriptional analysis of genes involved in acetate utilization in *Neurospora crassa*. *Molecular Microbiology* **2**, 599-606.

Tien, M. & Tu, C. -P. D. (1987). Cloning and sequencing of a cDNA for a ligninase from *Phanerochaete chrysosporium*. *Nature* **326**, 520-523.

Timberlake, W. E. & Barnard, E. C. (1981). Organization of a gene cluster expressed specifically in the asexual spores of *Aspergillus nidulans*. *Cell* **26**, 29-37.

Timberlake, W. E., Boylan, M. T., Cooley, M. B., Mirabito, P. M., O'Hara, E. B. & Willett, C. E. (1985). Rapid identification of mutation-complementing restriction fragments from *Aspergillus nidulans* cosmids. *Experimental Mycology* **9**, 351-355.

Ullrich, R. C., Novotny, C. P., Specht, C. A., Froeliger, E. H. & Munoz-Rivas, A. M. (1985). Transforming basidiomycetes. In *Molecular Genetics of Filamentous Fungi*, (ed. W. E. Timberlake), pp.39-57. Alan R. Liss, Inc.: New York.

Van Hartingsveld, W., Mattern, I. E., Van Zeijl, C. M. J., Pouwels, P. H. & Van den Hondel, C. A. M. J. J. (1987). Development of a transformation system for *Aspergillus niger* based on the *pyr*G gene. *Molecular and General Genetics* **206**, 71-75.

Vapnek, D., Hautala, J. A., Jacobson, J. W., Giles, N. H. & Kushner, S. R. (1977). Expression in *Escherichia coli* K12 of the structural gene for catabolic dehydroquinase of *Neurospora crassa*. *Proceedings of the National Academy of Sciences, USA* **74**, 3508-3512.

Viebrock, A., Perz, A. & Sebald, W. (1982). The imported pre-protein of the proteolipid subunit of the mitochondrial ATP synthase from *Neurospora crassa*. Molecular cloning and sequencing of the mRNA. *EMBO Journal* **1**, 565-571.

Vollmer, S. J. & Yanofsky, C. (1986). Efficient cloning of genes of *Neurospora crassa*. *Proceedings of the National Academy of Sciences, USA* **83**, 4869-4873.

Wahl, G. M., Lewis, K. A., Ruiz, J. C., Rothenberg, B., Zhao, J. & Evans, G. A. (1987). Novel cosmid vectors for genomic walking, restriction mapping and gene transfer. *Proceedings of the National Academy of Sciences, USA* **84**, 2160-2164.

Ward, M., Wilkinson, B. & Turner, G. (1986). Transformation of *Aspergillus nidulans* with a cloned, oligomycin-resistant ATP synthase subunit 9 gene. *Molecular and General Genetics* **202**, 265-270.

Ward, M., Wilson, L. J., Carmona, C. & Turner, G. (1988). The *oli*C3 gene of *Aspergillus niger*: isolation, sequence and use as a selectable marker for transformation. *Current Genetics* **14**, 37-42.

Weigel, B. J., Burgett, S. G., Chen, V. J., Skatrud, P. L., Frolik, C. A., Queener, S. W. & Ingolia, T. D. (1988). Cloning and expression in *Escherichia coli* of isopenicillin N synthetase genes from *Streptomyces lipmanii* and *Aspergillus nidulans*. *Journal of Bacteriology* **170**, 3817-3826.

Weltring, K. M., Turgeon, B. G., Yoder, O. C. & Van Etten, H. D. (1988). Isolation of a phytoalexin-detoxification gene from the plant pathogenic fungus *Nectria haematococca* by detecting its expression in *Aspergillus nidulans*. *Genetics* **68**, 335-344.

Yelton, M. M., Hamer, J. E., De Souza, E. R., Mullaney, E. J. & Timberlake, W. E. (1983). Developmental regulation of the *Aspergillus nidulans trp*C gene. *Proceedings of the National Academy of Sciences, USA* **80**, 7576-7580.

Yelton, M. M., Timberlake, W. E. & Van den Hondel, C. A. M. J. J. (1985). A cosmid for selecting genes by complementation in *Aspergillus nidulans*: selection for the developmentally-regulated *y*A locus. *Proceedings of the National Academy of Sciences, USA* **82**, 834-838.

Chapter 3

Novel methods of DNA transfer

J. W. Watts & N. J. Stacey

Over the past 40 years there has been a considerable effort to develop and expand methods for transfection. Many of the older procedures are very specialised and of limited application, only that which makes use of polyethylene glycol (Hinnen, Hicks & Fink, 1978; Dawson *et al.* 1978; Chang & Cohen, 1979; Negrutiu *et al.*, 1987) appears to be generally applicable, and even then principally to unwalled cells and protoplasts. The wall presents a very real obstacle to the entrance of macromolecules, but the precise reason why it is a barrier is not always clear, and much effort has been devoted to removing it, to produce protoplasts. The wall is, however, a structure of profound significance; its function is not simply mechanical and physical protection and support, but in most cases it possesses extremely complex and subtle functions associated with cell-cell signalling, differentiation and response to external stimuli (Varner & Lin, 1989). Wall removal and regeneration cannot, therefore, be thought of as analogous to taking the packing off an object and then replacing it; pathways of organization are disrupted and in the case of interdependent tissues, as in plants or fungal hyphae, total pathways of differentiation may be disturbed with profound consequences. Protoplasts may not, in fact, be too disorganised to regenerate walls and divide, but even with plant protoplasts, the new walls are quite atypical (Burgess & Linstead, 1976). Transfection through the cell wall is thus a desirable objective because it would avoid these disruptions.

Recently, some rather unusual methods for DNA transfer have been developed. Of these, two in particular seem to have real promise as methods of wide, perhaps general, applicability, although their potentials have not been fully explored: the use of DNA-coated microprojectiles (Klein *et al.*, 1987; Christou, McCabe & Swain, 1988) and the use of electric fields, electroporation (Potter, 1988; Neumann *et al.*, 1982). The apparatus needed for these techniques may appear exotic but is neither expensive nor difficult to construct. The technical problems associated with the methods are well worth examination to dispel some of the mystique surrounding them and to indicate directions in which the methods might proceed in the future. We consider first the use of the particle gun which avoids all the trauma of wall modification or removal and is, in effect, a method of micro-injection. We will then discuss electroporation which has already been used to transform walled bacteria (MacNeil, 1987; Chassy, Mercenier & Flickinger, 1988) and may ultimately be capable of routinely transferring DNA through plant cell walls.

Particle guns

General principles

Two types of particle gun have been described in some detail in the literature, one powered by chemical propellants (Klein *et al.*, 1987), the other by an electrical discharge (Christou *et al.*, 1988). The basic principle of action depends on acceleration of a macro-projectile, the leading face of which is loaded with micro-projectiles which have been coated with DNA; after reaching a high velocity the macro-projectile is halted by a stopper plate but the micro-projectiles are able to continue through holes in the plate and strike the sample. The material used for the micro-projectiles is either gold or tungsten, both characterized by high density and low chemical reactivity.

The gun developed by Sanford and his associates (Sanford *et al.*, 1987; Klein *et al.*, 1987) uses a chemical propellant, typically a 0·22 calibre blank cartridge, to accelerate a bullet-like polyethylene or nylon macro-projectile to velocities around 500 m s^{-1}. The DNA-coated micro-projectiles are suspended in a small drop of liquid in a recess on the leading face of the macro-projectile. The whole apparatus is evacuated to reduce problems from expanding gases and to reduce friction by air resistance to the passage of the micro-projectiles. This design of gun has proved very successful for transfection of single cells of higher plants (Klein *et al.*, 1988, 1989), including pollen (Twell *et al.*, 1989), mitochondria in fungi (Johnston *et al.*, 1988), and chloroplasts in algae (Boynton *et al.*, 1988). It has proved to be of little value for tissue transformation because of the limited number of 'hits' obtained with it and consequent production of chimaeras.

The electric gun (Christou *et al*,. 1988) is formed from two electrodes at the closed end of a short barrel. A small (5-20 μl) water droplet bridges the electrodes and discharge of a suitable charged capacitor produces an arc through the water and as a consequence a shock wave. The nature of the 'explosion' is not clear. The simplest explanation is explosive evaporation of the water but the effect is always accompanied by an arc, the behaviour of which is obscure but is associated with violent shock waves (Graneau, 1985). The micro-projectiles are gold particles coated with DNA. They are dried from suspension onto a disc of mylar film which is placed on the mouth of the gun. The shock wave from the discharge propels the disc upwards until it is stopped by a metal gauze. The micro-projectiles continue upwards and strike the specimen, which is usually contained in an inverted petri dish. As with the other gun, the operation

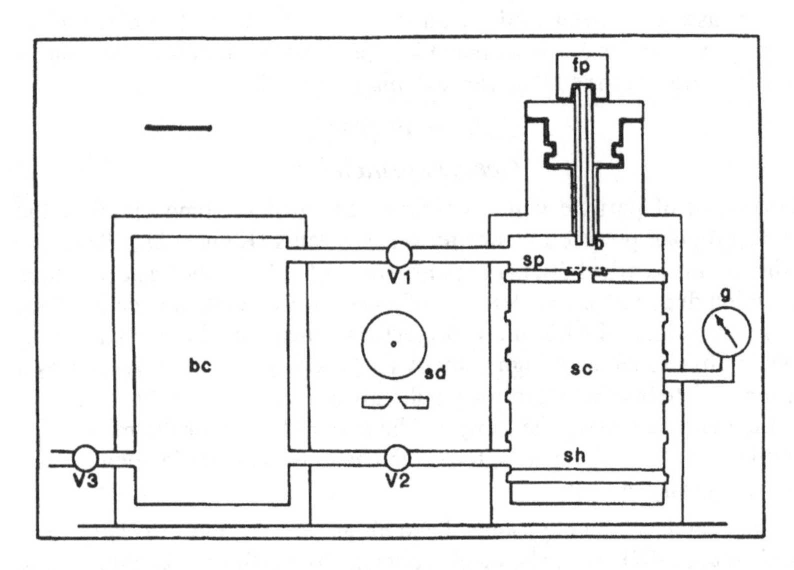

Fig. 3.1. Cartridge gun. The macro-projectile and blank cartridge are loaded into the breach of the barrel (b) under the firing pin (fp). The sample is placed on the sample holder (sh). The stopper plate (sp) holds the stopper disc (sd) (not to scale). The ballast chamber (bc) and sample chamber (sc) are evacuated with all three valves (V1-V3) open. When the gauge reads 70 to 72 cm Hg valves V2 and V3 are closed and the gun is fired. Vacuum is released by an air-inlet tap (not shown). Scale bar = 5 cm.

is performed in a partial vacuum. Despite its simplicity this gun has been used to transform soya cells to produce transformed plants (McCabe *et al.*, 1988). Unlike the cartridge gun it is not available commercially but is relatively easy to construct.

Design considerations

Cartridge gun

This has been described in some detail (Sanford *et al.*, 1987) and can be obtained (leased) commercially (Fig. 3.1). A detailed description is therefore out of place but some comments seem in order. The device is potentially dangerous and 'amateur' attempts to construct one should be attempted only in consultation with a gun-smith. Its use in the U.K. also requires a firearms certificate. The breach, firing mechanism, barrel and expansion chamber must be strong enough to contain the explosive force of the cartridge. The apparatus must also be vacuum-tight. Problems are experienced in controlling the gas exhausts. Propellants are chosen to give the correct burning rates and accelerations otherwise large amounts of ungasified material may be ejected into the sample chamber and the violence of the discharge may prevent efficient functioning of the stopper plate. A wide selection of blank cartridges is available, including those for nail guns and starting pistols. A range of charge sizes is also available, but the nature of the propellant is important; nitro loaded blanks seem most suitable, other types of charge may be unacceptably dirty. Problems may also be experienced with variability in the force of discharge produced by blanks since they are probably not subjected to the degree of control of the charge that is required for live ammunition. The stopper plate is intended not only to stop the macro-projectile but also to seal the sample chamber from the gases; consequently, there is a disposable disc, the stopper disc, at the centre of the steel stopper plate. The material from which the stopper disc (polycarbonate) and the macroprojectile (nylon or high molecular weight polyethylene) are made, are chosen so that they will fuse into one another on impact. This is unlikely to be more than partially effective, however. The macro-projectile will not prevent gas leakage past it during firing and the plastic stopper disc will distort on impact permitting gas escape around its edges. The ballast chamber is an attempt to buffer the system against some of these problems and small holes are also drilled in the muzzle of the barrel to allow lateral gas escape. The violence of the gas movement, though, is such that all these modifications are palliatives, not cures. A 'dirty' charge can very easily cover the sample with combustion debris and inhibit subsequent growth.

Loading of the DNA is a problem which has not been satisfactorily resolved. The usual method entails precipitating DNA onto tungsten particles by addition of spermidine and $CaCl_2$ to give a suspension which is loaded into a recess in the leading face of the macro-projectile (Klein *et al.*, 1987). Tungsten particles, although relatively cheap, are very prone to aggregation (Sanford, 1988), particularly when the powder has aged. There is also a pronounced tendency for the DNA-tungsten mixture to aggregate and although sonication immediately before use reduces this problem, there remain aggregates which are not suitable for transferring DNA into cells. Workers in our Institute have observed that the transformed zones in callus often appear to lie at the base of craters, as though a great deal of cell destruction, perhaps due to liquid micro-droplets or to aggregated particles, had been required to produce a few good hits. The original papers specified a tungsten : DNA (w/w) ratio of about 1000 : 1 but more recent work (Daniell *et al.*, 1990) has used a ratio much closer to 1 : 1, with correspondingly reduced amounts of tungsten particles in each discharge. Tungsten carbide may be considered as an alternative to tungsten. It is a brittle non-metal some 10% less dense than tungsten, but can be crushed readily in a small ball-mill to give suitable particle sizes.

These have the great advantage that they do not have the same tendency as metal powders to aggregate and can be stored indefinitely as aqueous suspensions. The method of sample loading used with the electric gun (Christou *et al.*, 1988) has some advantages but cannot easily be used with the cartridge gun.

One aspect of guns that should be kept in mind is the need to work with the apparatus evacuated, to reduce air resistance to the motion of the free micro-projectiles, and to control shock waves from the propellant. In the case of the cartridge gun around 28 in (71 cm) vacuum is recommended. This may not be acceptable to the tissue.

Electric gun

The basis of propulsion in the electric gun is the production of a shock wave by discharge of a capacitor through water. The original papers (Christou *et al.*, 1988; McCabe *et al.*, 1988) describe the use of a 2 μF capacitor charged to 14 kV, a potentially lethal and rather expensive specialist device. The energy delivered by such a device is of the order of 200 J but a high voltage is essential to produce the arc explosion. Simple discharge of 200 J through water without an arc has no such effect (Graneau, 1985). It is possible to reduce the voltage very considerably and still produce an arc, provided the capacitor size is increased to hold sufficient energy. The micro-particles are made of gold, which has less tendency than tungsten to clump, coated with DNA by drying down particles and DNA solution together in a stream of nitrogen. Whilst gold particles can be obtained commercially it is useful to bear in mind that particles in a range of sizes can be produced by reducing solutions of gold chloride. Particles of 0.5 to 1.0 μm can be prepared by addition of a few drops of hydrogen peroxide to a slightly alkaline solution of gold chloride (1 mg ml^{-1}) and can be inhibited from clumping by addition of a suitable stabiliser, e.g. protamine (0.1 mg ml^{-1}), before washing. The coated particles are taken into suspension in ethanol and dried down onto plastic discs, the macro-projectiles. The choice of plastic for the discs is important and mylar film is recommended. The impact of the shock wave can tear weaker films, particularly when more powerful discharges are used.

The construction of a capacitor of the type required is relatively straightforward. If the working voltage is limited to 3 kV simple capacitors can be purchased, although these are rather expensive. We have found it simpler to construct composite capacitors using 400 V DC working electrolytic capacitors. With suitable precautions these can be arrayed in series and in parallel to provide a composite capacitor of almost any desired value. It must be remembered, however, that the internal impedance of an electrolytic capacitor is substantially greater than that of a similar sized simple non-polar metal-film capacitor, so that the rate at

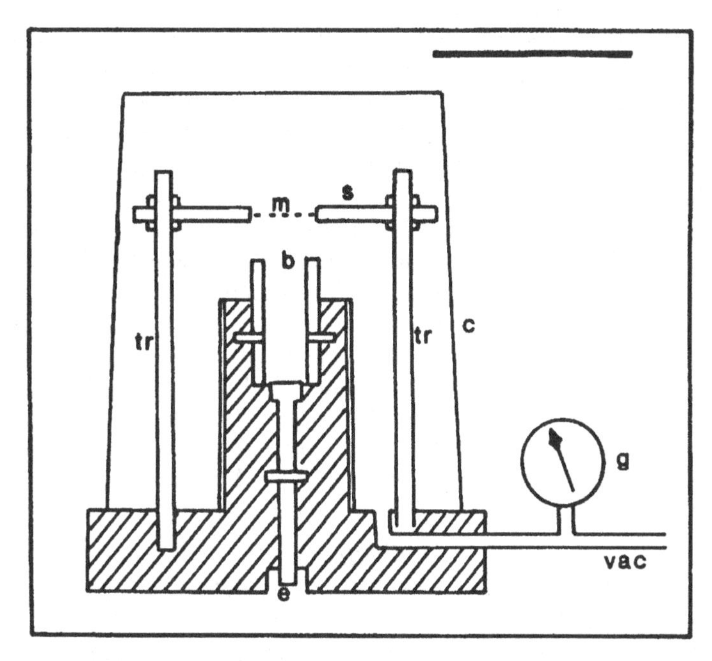

Fig. 3.2. Electric powered gun. A stainless steel barrel (b) and central electrode (e) are held firmly by cross-pinning in a plastic block (shaded). The stopper plate (s) has a central zone of wire mesh (m), the height of which is adjustable on threaded rods (tr). The sample, in an inverted Petri dish, rests on the stopper plate. The whole apparatus is sealed, with a plastic cover (c). To operate, a drop of water (5 to 50 μl) is placed at the bottom of the barrel, bridging the electrodes. The micro-projectiles on a mylar plastic disc rest on the mouth of the barrel. The apparatus is evacuated to 60 to 65 cm Hg and the capacitor is discharged across the electrodes, the shock wave throwing the disc against the wire mesh. Scale bar = 5 cm.

which energy can be delivered to the arc is correspondingly slower. The detailed construction of the gun itself has been described by Christou *et al.* (1988). Fig. 3.2 shows a design we have found reliable in use and able to take very much higher powers than specified in the original papers. The stainless steel barrel and inner electrode must be firmly cross-pinned into the plastic surround to prevent movement during repeated discharges.

There is nothing fundamentally new about the design and construction of electric guns. Rail guns, which use a metal arc plasma as a propellant, have long been an area of study whilst very powerful water guns, similar in design to the electric particle gun, have been described (Frungel, 1965; Graneau, 1985, 1989). Transfer of energy by a shock wave in air in a partial vacuum is inevitably very inefficient. Energy-transfer in the true water gun is achieved by striking an arc at the base of a column of water. The expansion of the arc plasma is then transferred directly to the column of water which is violently accelerated to high velocities. A macro-projectile resting on the top of the incompressible column of water is similarly accelerated. The design in Fig. 3.2 is readily modified to permit the use of bullet-shaped macro-projectiles in much the same way as the cartridge gun. Energy transfer to the macro-projectile is very inefficient; a survey of some of the published data suggests that even in the more efficient guns, and the gun described above is not one of these, only about 1% of the energy in the capacitor is translated into motion of the macro-projectile, the rest is lost elsewhere in the apparatus. It follows that to produce velocities comparable with those obtained with chemical propellants very large capacitors will be needed. For example, if the mass of the macro-projectile is 0·1 g and the desired velocity 500 m s^{-1}, calculation indicates that a capacitor of about 300 μF at 3 kV is needed. The circuitry to deliver such a charge with some efficiency would need to be correspondingly more robust.

Applications

The great advantage of the particle gun is its ability to transfect walled cells. The cartridge gun has been used to transform a wide range of individual cells. Particles in the cartridge gun have velocities of the order of 500 m s^{-1} and are able to penetrate the cell wall and, in favourable circumstances, pass through several cell layers. Typical of the results obtained are transient expression in a range of plant tissue including pollen, using pollen-specific promoters (Twell *et al.*, 1989), and chloroplasts of tobacco cells (Daniell *et al.*, 1990), and transformation of callus, mesophyll cells (Klein *et al.*, 1988), fungi (Johnston *et al.*, 1988; Armaleo *et al.*, 1990) and chloroplasts of *Chlamydomonas* (Boynton *et al.*, 1988). Recently, the production of transformed cotton by bombardment of embryogenic callus has been reported (Finer & McMullen, 1990).

Initially, one of the most encouraging results with the electric gun was its use to transform soya bean callus and soya apical meristem cells from which transformed plants were recovered (Christou *et al.*, 1988; McCabe *et al.*, 1988). There have also been uncorroborated reports of transformation of maize pollen. The unusual feature of these successes is that they have been achieved using particle velocities at least one order of magnitude less than those obtained with the cartridge gun. Simple experiments with the electric gun show that particle velocities are unlikely to be greater than a few tens of metres per second. The velocity of sound in air is about 330 m s^{-1} and the shock wave from the electric discharge is unlikely to be travelling faster than this when it strikes the plastic disc supporting the DNA-coated gold particles. Furthermore, the shock wave will possess relatively little energy, particularly since it propagates in a partial vacuum, and energy-transfer to the plastic disc will be inefficient. It is possible to modify the gun and increase its power to obtain greater velocities, though the use of bullet-like macro-projectiles may then be desirable. However, the original successes were obtained without these modifications and with relatively low velocities. Why is this so when protagonists of the cartridge gun consider that much higher velocities are essential? It is possible that the choice of tissue with the electric gun has been the critical factor. Pollen and meristematic material are highly cytoplasmic with small or negligible vacuoles. Plasmolysis, with shrinking of the protoplast from the cell wall may thus not be a serious problem. Provided the cell is in turgor the wall will be penetrated readily because its elastic properties will not be able to absorb the impact of the micro-projectile. Furthermore, walls of meristematic tissue are extremely thin while the exine layer of the pollen grain may well be too brittle to absorb the impact and may therefore not constitute the barrier its chemical stability might suggest.

Future prospects

The attraction of particle guns is their ability to penetrate walls and to carry DNA directly into cell organelles like chloroplasts and mitochondria (Boynton *et al.*, 1988: Johnston *et al.*, 1988). Guns may provide a valuable means for transforming fungi without the need to make protoplasts or the trauma of chemical and electrical methods. For higher plants, the transfection of pollen may offer a way to by-pass problems of tissue culture. Micro-injection of pollen has been attempted by many groups with signal lack of success; the use of guns may be the answer. In many cases, however, there seem to be no significant advantages to outweigh the problems. Electroporation has been highly successful in transforming bacteria (MacNeil, 1987; Chassy & Flickinger, 1987) and there are alternative chemical approaches (Chang & Cohen, 1979; Klebe *et al.*, 1983). Thus the use of guns to transfect bacteria may be of academic interest only; in particular, the acceleration of sufficiently small particles may be a problem.

Electroporation

General principles

When a brief voltage pulse is used to rupture the plasma membrane, the protoplast remains permeable to small molecules for some minutes (Zimmermann & Vienken, 1982; Lindsey & Jones, 1987). Molecules present in the medium, provided they can pass through the cell wall and overcome any electrostatic repulsions, may then be able to enter the cell, particularly at the moment the pulse is applied. This process of permeabilization, or electroporation as it is usually termed, has proved to have wide applications in introducing macromolecules, including DNA, as well as small molecules into cells or protoplasts of bacteria, fungi, plants and animals. In simple terms the process has been visualised as formation of pores in the plasma membrane, but this view does not provide an entirely satisfactory explanation for the experimental observations and to some extent the potentials and limitations of the method must be established empirically.

The plasma membrane is, at a very simple first approximation, a lipid bilayer with very strong electrical insulating properties. The breakdown voltage for a bilayer of this type is about 1 V (Zimmermann & Vienken, 1982); thus, in the case of a fungal protoplast with a diameter 2 to 5 μm, a field strength of 4 to 10 kV cm^{-1} will be required to disrupt the membrane. The corresponding fields for plant cells (10 to 40 μm diameter) and bacteria (1 μm diameter) are 0.5 to 2 and 20 kV cm^{-1} respectively. The early workers (Neumann *et al.*, 1982; Shivarova *et al.*, 1983), who used brief pulses of a few μs duration, employed very high field strengths to induce electroporation, but later workers (Potter, Weir & Leder, 1984) found that much lower field strengths, of the orders given above, were effective. The first power pack we used, to inoculate tobacco mesophyll protoplasts with viral RNA, generated up to 16 kV, giving notional, but quite unusable fields of 40 kV cm^{-1} across a 4 mm inter-electrode spacing (Watts, King & Stacey, 1987). Since we were working with plant protoplasts with a diameter of 25-30 μm, in practice fields of only 1 to 2 kV cm^{-1} were effective.

When the plasma membrane is subjected to relatively much lower fields another type of electrical breakdown occurs. This is a slower process than that caused by high field strengths, taking place in ms rather than μs. The plasma membrane is in reality far from being a simple lipid bilayer. It is studded with proteins and associated polysaccharides which may account for more than 50% of the surface area (Jain, 1988). These components form the pores and ion pumps which provide the gateways by which nutrients and other molecules enter or leave the cell, as well as fulfilling more obscure functions, like transmembrane signal transduction. It is possible that this slow depolarisation is a consequence of the presence of these non-lipid components in the plasma membrane.

The action of ion pumps within the membrane produces a potential difference between the cytosol and the external medium, usually in the range 10 to 100 mV (Jain, 1988). The protoplast thus has a net negative charge, moving to the anode in an electric field and repelling negatively charged material. The repulsion of negatively charged objects is a major problem in attempts to transfer nucleic acids into the cell and many of the procedures used in transfection have the elimination of this repulsion as one of their purposes. For example, one of the roles of polycations like poly-L-ornithine in the infection of plant protoplasts with viral nucleic acids is to modify electrical charges on the nucleic acids and cell membrane to permit close approach (Motoyoshi, Watts & Bancroft, 1974). Similarly, one of the functions of PEG is probably to aggregate and precipitate nucleic acids and viruses onto the plasma membrane (Hebert, 1963). The remarkable feature of electroporation is the efficiency with which these repulsive forces are overcome. Exactly how this occurs is not clear. The simple model of electroporation assumes that the electrical pulse creates small holes in the membrane, of about 5 nm diameter, and the nucleic acids diffuse through these (Zimmermann & Vienken, 1982). In our experience, this does not account for experimental results. When electroporation of small molecules is studied, the membrane is found to remain permeable for some minutes after the pulse, the duration of the effect being greater at 4°C than at room temperature (Zimmermann & Vienken, 1982). We observe, however, that nucleic acids must be present during the voltage pulse for efficient transfection to occur (Watts *et al.*, 1987). It seems much more probable that for negatively charged material a definite driving force may be required to force macromolecules through the electrically induced pores. Positively charged material associates spontaneously with the plasma membrane and is much more readily taken into the cell during electroporation. This may be due to activation of the membrane by the voltage pulse rather than to formation of 'pores' (Watts *et al.*, 1987). It is known, for example, that when passively inoculating plant protoplasts with the assistance of polycations, the character of the protoplast surface is critical and best results are obtained by washing the protoplasts immediately before inoculation (Otsuki *et al.*, 1972). Activation is a very transient phenomenon and may have some affinities with that observed with electroporation.

Design considerations

The basic apparatus for electroporation, and one that gives excellent service, is a simple capacitor with a voltage source to charge it and a switch to disconnect the power supply and connect the capacitor to the electroporation chamber. A conventional electrophoresis power supply giving up to 3 kV is a satisfactory voltage source. Switching may be done with relays. Solid state switching may be used if precautions are taken to limit current, to prevent destruction of the switching circuitry. A pulse limiter in the form of a suitable resistance, value around 100 Ω, may therefore be desirable. The capacitor may present problems if high values at high working voltages are required. A 50 nF capacitor capable of working voltages up to 10 kV has given very good results for transfection of RNA into plant protoplasts (Watts *et al.*, 1987). For some types of work, particularly transient expression and transformation with DNA, better results may be obtained with lower voltages and larger capacitors. It is not usually necessary to use fields above 1 kV cm^{-1} with cells of higher organisms but much higher voltages and relatively large (25 μF) capacitors have been required for electroporation into the smaller cells of fungi and bacteria (Hashimoto *et al.*, 1985; MacNeil, 1987; Chassy & Flickinger, 1987; Meilhoc, Masson & Teissié, 1990).

The insulating properties of air are finite and insulation breakdown and arcing through the air may occur at field strengths much above 5 kV cm^{-1}, particularly if the electrodes have sharp, exposed edges. An electric arc has a very low resistance so arcing causes rapid discharge of the capacitor with current flow which may destroy the circuitry. With arcing, the voltage drop across the cuvette will be too low for electroporation but the sample will probably be ejected explosively from the cuvette and the cuvette itself may disintegrate. It follows that field strengths much above this are unrealistic without special precautions to prevent arcing and the usual working maximum is around 7·5 kV cm^{-1}. If the inter-electrode distance is 4 mm, a maximum supply voltage of 3 kV is required. Capacitors of several μF with this working voltage can be purchased and reliable electroporation equipment incorporating them is available commercially. It is, however, very simple to construct composite capacitors of the type used with electric particle guns and these can then be used both for electroporation and particle gun work.

Simple capacitative discharge is satisfactory for many purposes but much better results can often be obtained by use of shaped pulses (Nishiguchi *et al.*, 1987; Hibi, 1989). A square pulse generator is essentially a large capacitor which is only partially discharged during the pulse. If too much current is drawn the pulse deviates unacceptably from a square wave form. Current must therefore be limited during discharge, although this is not usually a serious problem. However, consideration must be given, as with all discharges of large capacitors, to heating effects. The first square pulse generator we used delivered a maximum of 950 V for 25 ms at a maximum current of about 5 to 10 A. This corresponds to about 5 to 10 kW of power, which would in theory raise the temperature of a 1 ml sample by about 25°C. In practice, the maximum temperature rise would be rather less than this, but as a matter of common-sense samples should have as low a conductivity as possible to reduce both heating and deviations from a square wave form, particularly when using longer pulses. A useful feature of square pulses is that once conditions have been established for a particular interelectrode spacing, these can be applied to any chamber volume provided the electrode spacing is the same.

The construction of an electroporation chamber presents no difficulties. The original design of Potter *et al.* (1984) based on a disposable plastic cuvette is adequate and plastic electroporation cuvettes of a more substantial character can be purchased. We have concluded from experience, however, that it is usually an advantage to control temperature rather more effectively than is possible with plastic chambers and have constructed a chamber based on a standard 1 cm silica cuvette using aluminium electrodes of sufficient thickness (3 mm) to give an interelectrode gap of 4 mm. The cuvette is suspended in a water bath to control temperature. The use of silica or glass allows efficient temperature control whilst the thermal capacity of the chamber provides a larger buffer against heating effects when using higher energy discharges.

Electroporation at high field strengths is best performed in specially constructed cells. The electrodes should be of massive, polished metal, preferably water cooled, and the edges of the metal should not form part of the discharge zone. A simple design is block aluminium for the upper and lower electrodes with a sheet of polythene between to provide the interelectrode spacing. The sample chamber is a circular hole in the centre of the polythene sheet. The advantage of this design is simplicity and flexibility. The interelectrode distance can be reduced to less than 1 mm without difficulty so that very high field strengths are attainable with standard laboratory power supplies and expensive high voltage pulse generators are not necessary. Micro-chambers with adjustable inter-electrode distances have been described, the sample being held in place by capillarity (Galvin & Hanawalt, 1988).

A variety of media, some based on physiological salines (Potter *et al.* 1984), has been employed in electroporation, usually with some underlying rationale. In our experience, and in agreement with that of several other workers (Calvin & Hanawalt, 1988; McIntyre & Harlander, 1989) care is needed in choice of medium. A medium serves primarily to suspend the cells during electroporation and whilst it may seem desirable to make its composition similar to that of the cytosol, this must inevitably affect the form of the voltage discharge and even the electrical properties of the cells during electroporation., There is also a real danger of introducing ions which may be toxic into the cell. For example, the intracellular concentration of free calcium is very low and the presence of millimolar concentrations of calcium in a medium for electroporation may prove toxic (Watts *et al.*, 1987). Unless there is good experimental reason to do otherwise, the first choice should be a simple, electrically non-conducting medium, e.g. aqueous solutions of sucrose or mannitol. The osmolarity of the medium is usually most important, somewhere in the range 0·25 to 0·5 M is usually satisfactory.

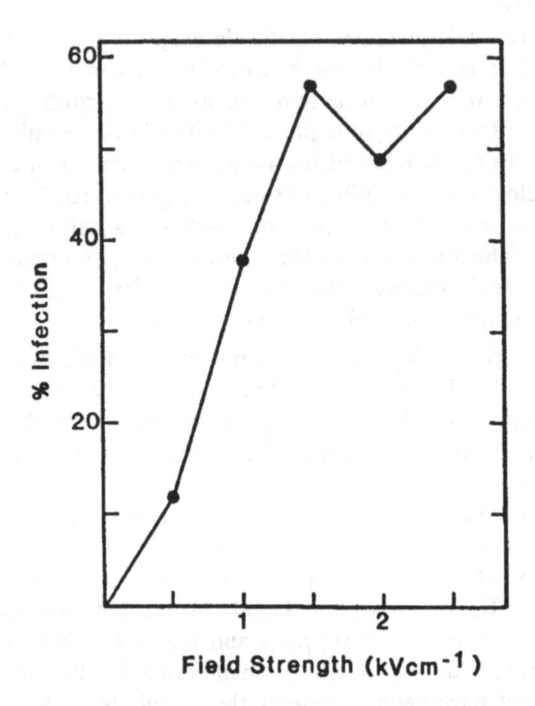

Fig 3.3. Relationship between field strength and infection with viral RNA during electroporation. Virus RNA (25 μg ml^{-1}) was electroporated into tobacco mesophyll protoplasts using a 50 nF capacitor. Percentage infection was determined after 40 h culture by fluorescent antibody staining.

Results

The first experiments with electroporation used very high field strengths to transform animal cells and bacterial protoplasts, conditions beyond the facilities of the normal laboratory (Neumann *et al.*, 1982; Shivarova *et al.*, 1983), and it was not until Potter *et al.* (1984) demonstrated electroporation using a simple chamber and power supply that the method was seriously exploited. A very substantial body of published work now exists in which the method has been used to inoculate animal and bacterial cells and plant protoplasts with a variety of macromolecules and particles. Fungi have been rather neglected. Karube, Tamiya & Matsuoka (1985) and Hashimoto *et al.* (1985) reported transformation at very low efficiencies of yeast spheroplasts and intact cells respectively, but Meilhoc *et al.* (1990) have recently described high frequency transformation of yeast cells. Rather than review the literature it will probably be more profitable to examine the underlying reasons for the different strategies.

Initially we used electroporation to inoculate plant protoplasts with viral RNA (Watts *et al.*, 1987). This approach has the advantage that successful inoculation can be scored as a percentage of inoculated cells by fluorescent antibody staining some 30 hours after electroporation. Brome mosaic virus (BMV) and cowpea chlorotic mottle virus (CCMV), are multicomponent viruses which have been well characterised. It is known that successful infection of tobacco protoplasts requires of the order of 1000 virus particles (Motoyoshi, Bancroft & Watts, 1973). This high number derives in part from the multi-component character of the virus and in part from its low specific infectivity. The infectivity of free viral RNA is similar. The method therefore provides a measure of the fraction of protoplasts that has taken up at least this amount of viral RNA. Using a 50 nF capacitor at different field strengths, equivalent to a pulse length around 10 μs, the results agreed well with the theoretical requirements of the simple model (Fig. 3.3). Electroporation has proved to be very much more efficient than any other known method of inoculation with RNA.

When the experiments were performed with intact virus however, a different picture emerged. The positively charged BMV infected readily under these conditions, but the negatively charged CCMV only poorly. A brief, earlier report of successful inoculation of protoplasts of *Vinca rosea* suspension culture cells with tobacco mosaic virus (TMV) (Okada, Nagata & Takebe, 1986) used rather different conditions. Like CCMV, TMV is a negatively charged virus and so is repelled by the plasma membrane; but in contrast to CCMV, it is a single component virus of high specific infectivity and was used at very high concentrations (500 μg ml^{-1}). In these experiments infection with TMV required large capacitors (ca 100 μF), i.e. longer pulses. The use of larger capacitors often causes an unacceptably high level of damage and is usually avoided. Nishiguchi *et al.*, (1987) showed that pulse duration was of significance with TMV and found that TMV could be efficiently inoculated into tobacco protoplasts by square pulses that were about 10 ms long. Thus electroporation of larger sized objects like viruses (ca 25 nm diameter) requires longer pulses. One of our longer term objectives is to electroporate organelles, and we therefore turned our attention to the use of square rather than exponentially decaying pulses.

Fig. 3.4 shows the results of an experiment with square pulses to study the relation between field strength, pulse duration, and percentage infection with BMV RNA. As in all the experiments with square pulses the medium was a simple non-conducting solution of 0·7 M mannitol. As a contrast, Fig. 3.5 shows the comparable data for intact CCMV particles. The difference in response is striking and demonstrates that electroporation of virus requires longer pulses. Both experiments show that the field

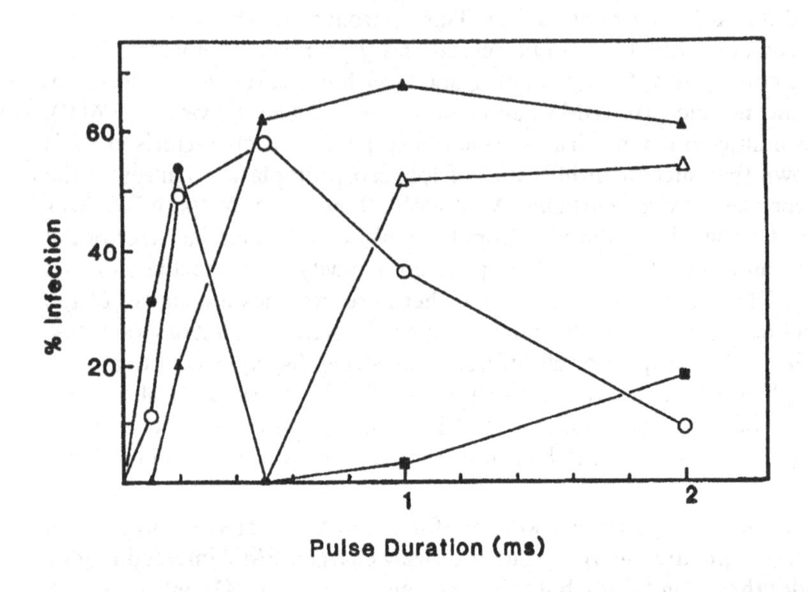

Fig. 3.4. Relationship between pulse duration, field strength and infection with BMV RNA during electroporation by square pulses. Percentage infection was measured after 40 h culture by fluorescent antibody staining. Closed circles, 2 kV cm^{-1}; open circles, 1·5 kV cm^{-1}; closed triangles, 1 kV cm^{-1}; open triangles, 0·5 kV cm^{-1}; closed squares, 0·25 kV cm^{-1}.

strength can be greatly reduced as the pulse duration increases. There is, however, a lower limit to effective field strength; about 100 V cm^{-1} in the case of tobacco mesophyll protoplasts, which is equivalent to about 150 mV across the plasma membrane. It appears, then, that electroporation may take place by two rather different mechanisms, but that larger particles obligatorily require much longer pulses to enter the cell.

A major application of electroporation is transfection with DNA to obtain either short-term transient expression or long-term transformation. Fig. 3.6 shows the relation between pulse duration and transient expression. Note that the result is similar to that obtained with virus particles, but not, we suppose, for the same reason. If short pulses (ca 1 ms) are used the level of transient expression is found to increase sharply as the field strength exceeds 1 kV cm^{-1}. The level of transient expression depends on the amount of DNA entering the cell, in contrast to infection

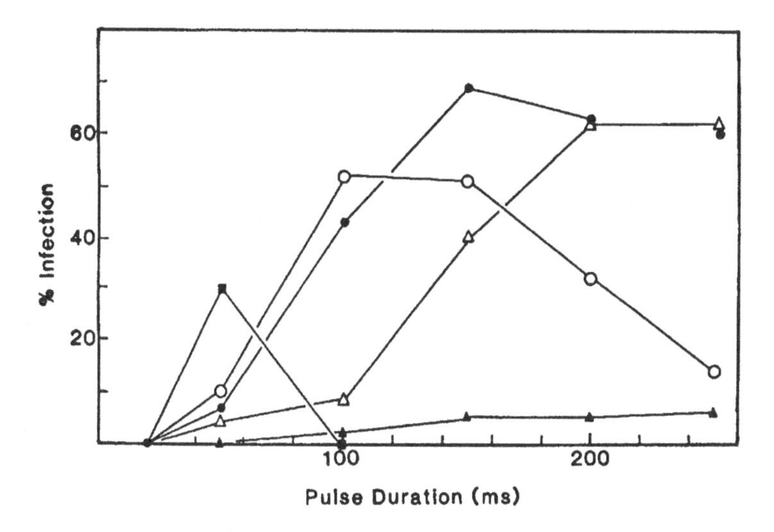

Fig. 3.5. Relationship between pulse duration, field strength and infection with intact CCMV particles during electroporation by square pulses. Closed squares, 1 kV cm^{-1}; open circles, 0·5 kV cm^{-1}; closed circles, 0·375 kV cm^{-1}; open triangles, 0·25 kV cm^{-1}; closed triangles, 0·125 kV cm^{-1}.

by viral RNA where the assay records only whether or not the cell is infected. The amount of DNA entering the cell will thus increase with the duration and intensity of the pulse, and the level of transient expression will increase similarly until saturation is reached. We are effectively studying electrophoresis into protoplasts.

The conditions used to electroporate bacteria are of some interest. In general, rather extreme conditions are required. Theoretically, electrical breakdown of the plasma membrane should occur at around 20 kV cm^{-1}, a field strength far in excess of that usually available. The many successful reports in the literature centre around a field strength of about 7 kV cm^{-1} and a pulse, usually provided by a 25 μF capacitor, with a duration of about 10 ms (Chassy *et al.*, 1988). It is probable, therefore, that the mechanism of electroporation depends on the breakdown of the plasma-membrane that occurs at lower voltages. The possibility of heat shock

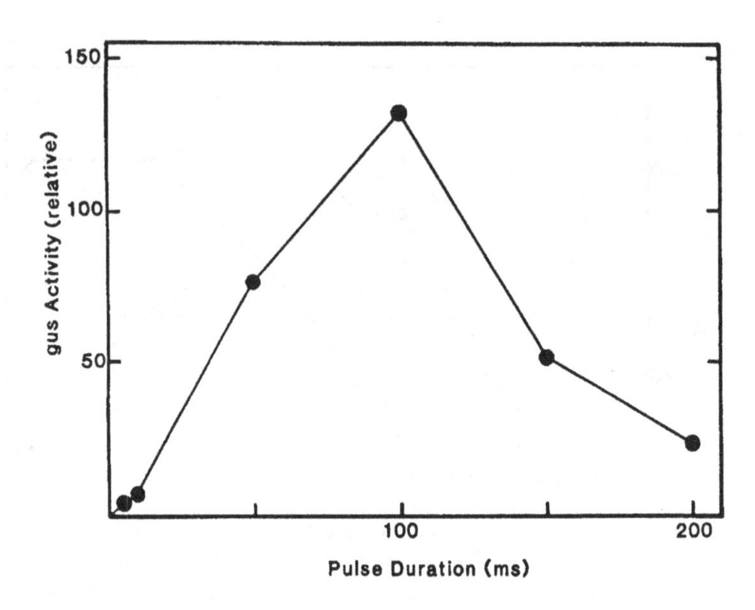

Fig. 3.6. Effect of pulse duration on transient expression. Tobacco mesophyll protoplasts were electroporated with DNA (pMJD 67, a construct containing the *gus* gene) (25 μg ml^{-1}) at 0·5 kV cm^{-1} using square pulses of different durations. Transient expression (*gus* activity) was measured 40 h later.

must be anticipated in the extreme conditions used when electroporating intact bacterial cells. Some 50 to 100 J of energy may be dumped into the system when the pulse is applied, although only a part of this will be into the sample itself. The sample volume is usually 0·5 ml, and may be considerably less, so in theory a temperature rise of 100°C is possible. It is by no means clear what the relative contributions of the voltage pulse and thermal shock may be to transformation and to the considerable number of cell deaths often observed during electroporation.

The recent work of Meilhoc *et al.* (1990) in which yeast cells were transformed with high efficiency is of considerable interest. These workers used square pulses to electroporate DNA through the walls and plasma membrane and present results that closely parallel our work with square pulses and complement the data on electroporation of bacteria.

Table 3.1. Electroporation with shaped pulses. Tobacco mesophyll protoplasts were electroporated with DNA (pMJD 67) (20 μg ml^{-1}) using the square pulses and durations shown, and assayed for *gus* activity 40 h later.

First Pulse		Second Pulse		
Field strength ($kV\ cm^{-1}$)	Duration (ms)	Field strength ($kV\ cm^{-1}$)	Duration (sec)	Relative *gus* activity
0·0	0	0	0	22
0·5	1	0	0	31
0·75	1	0	0	24
1·0	1	0	0	24
0·0	0	0·12	2	130
0·5	1	0·12	2	620
0·75	1	0·12	2	750
1·0	1	0·12	2	2000

Since there are two mechanisms by which the plasma membrane can be gated in electroporation, that is, high voltage with brief pulses or low voltages with long pulses, it is of interest to explore the effects of shaped pulses to optimize efficiency of transfection and to reduce cell destruction. The pulse shape we have been exploring consists of a brief (1 to 5 ms) high voltage square pulse followed by a long (seconds) low voltage pulse. The result of an experiment assaying transient expression in plant protoplasts is shown in Table 3.1. The combination of pulses appears significantly more effective than either of the two pulses alone, exploiting the ability of the brief high voltage pulse to rupture the membrane and then for the long low voltage pulse to both keep the membrane gated and to drive substantial amounts of DNA into the cell.

The field of electroporation abounds in paradoxes. Almost all the work with plants has involved protoplasts, little work has been published on fungi, and yet a substantial body of work on bacteria has used intact organisms. The possibility of electroporation with walled cells has enormous attractions for plant workers since it would allow transfection of DNA into intact tissues (Morikawa *et al*., 1986; Ahokas, 1989). This particular area has been engaging our attention recently, with some success.

Although there is a great diversity of material in the walls of bacteria, fungi and plants, there is also an underlying unity (Bartnicki-Garcia, 1968; Hancock & Poxton, 1988; Varner & Lin, 1989). In general, the wall is constructed from a lattice of strongly cross-bonded polymers. In some organisms the wall may effectively be a three-dimensional, covalently linked structure whilst in others much of the integrity may depend on hydrogen bonds. Since metabolites and even macromolecules like secretory enzymes must be able to enter or leave the cell, the wall lattice has a relatively open structure, although the openings may be filled with other, looser components, For example, the primary wall of a plant cell is often a lattice of cellulose microfibrils cross-bonded by hemicellulose with the interstices filled with a pectin gel (Varner & Lin, 1989).

Two obstacles must therefore be overcome: the DNA must be able to pass through the interstices and the plasma membrane must remain open long enough for the DNA to cross the wall and enter the cytosol. It transpires that the long pulses we have been using for work with intact virus and to optimise transient expression with DNA do enable viral RNA to cross the wall of mesophyll protoplasts and cause infection. Additionally, it is possible to drive DNA through the wall and give transient expression. In both these cases we have used cells with walls that have been modified by exposure to enzymes, but this does not seem strictly necessary. The work with bacteria shows that wall modification may sometimes be desirable (Scott & Rood, 1989) but the major constraint appears rather to be the physiological state of the cell (Miller *et al.*, 1988; Fiedler & Wirth, 1988). It seems likely that using carefully shaped pulses these successes may be extended and placed on a firm footing.

Future prospects

Electroporation is already a versatile technique for transfecting intact bacteria and yeasts and a range of wall-less cells and protoplasts. There are good reasons to assume that the technique will be applicable to a wide range of walled cells once conditions have been explored. This in turn may allow transformation of cells in organized tissues.

Conclusions

Particle guns and electroporation have proved of considerable value in the transformation of a wide range of cell types. Neither method should be seen as a substitute for well-tried conventional methods but rather as useful alternatives, particularly for difficult subjects. Much development and refinement will be needed before the particle gun can be fully exploited and a more powerful electric gun could usefully replace the cartridge powered gun. Electroporation may in the end, however, prove to be the versatile general procedure that most workers seek.

References

Ahokas, H. (1989). Transfection of germinating barley seed electrophoretically with exogenous DNA. *Theoretical and Applied Genetics* **77**, 469-472.

Armaleo, D., Ye, G. -N., Klein, T. M., Shark, K. B., Sanford, J. C. & Johnston, S. A. (1990). Biolistic nuclear transformation of *Saccharomyces cerevisiae* and other fungi. *Current Genetics* **17**, 97-103.

Bartnicki-Garcia, S. (1968). Cell wall chemistry, morphogenesis, and taxonomy of fungi. *Annual Review of Microbiology* **22**, 87-108.

Boynton, J. E., Gillham, N. W., Harris, E. H., Hosler, J. P., Johnson, A. M., Jones, A. R., Randolph-Anderson, B. L., Robertson, D., Klein, T. M., Shark, K. B. & Sanford, J. C. (1988). Chloroplast transformation in *Chlamydomonas* with high velocity microprojectiles. *Science* **20**, 1534-1538.

Burgess, J. & Linstead, P. J. (1976). Scanning electron microscopy of cell wall formation around isolated plant protoplasts. *Planta* **131**, 173-178.

Calvin, N. M. & Hanawalt, P. C. (1988). High efficiency transformation of bacterial cells by electroporation. *Journal of Bacteriology* **170**, 2796-2801.

Chang, S. & Cohen, S. N. (1979). High frequency transformation of *Bacillus subtilis* protoplasts by plasmid DNA. *Molecular and General Genetics* **168**, 111-115.

Chassy, B. M. & Flickinger, J. L. (1987). Transformation of *Lactobacillus casei* by electroporation. *FEMS Microbiology Letters* **44**, 173-177.

Chassy, B. M., Mercenier, A. & Flickinger, J. (1988). Transformation of bacteria by electroporation. *Tibtech* **6**, 303-309.

Christou, P., McCabe, D. E. & Swain, W. F. (1988). Stable transformation of soybean callus by DNA-coated gold particles. *Plant Physiology* **87**, 671-674.

Daniell, H., Vivekananda, J., Nielsen, B. L., Ye, G. N., Tewari, K. K. & Sanford, J. C. (1990). Transient foreign gene expression in chloroplasts of cultured tobacco cells after biolistic delivery of chloroplast vectors. *Proceedings of the National Academy of Sciences, USA* **87**, 88-92.

Dawson, J. R. O., Dickerson, P. E., King, J. M., Sakai, F., Trim, A. R. H. & Watts, J. W. (1978). Improved methods for infection of plant protoplasts with viral ribonucleic acid. *Zeitschrift für Naturforschung* **33C**, 548-551.

Fiedler, S. & Wirth, R. (1988). Transformation of bacteria with plasmid DNA by electroporation. *Analytical Biochemistry* **170**, 38-44.

Finer, J. J. & McMullen, M. D. (1990). Transformation of cotton (*Gossypium hirsutum* L.) via particle bombardment. *Plant Cell Reports* **8**, 586-589.

Frungel, F. (1965). *High Speed Pulse Technology*, Vol. 1. Academic Press: New York.

Graneau, P. (1985), *Ampère-Neumann Electrodynamics of Metals*. Hadronic Press Inc.: Nonantum, Mass., USA.

Graneau, P. (1989). Alpha-torque forces. *Electronics World and Wireless World* **95**, 556-559.

Hancock, I. & Poxton, I. (1988). *Bacterial Cell Surface Techniques*. John Wiley & Sons Ltd.: New York.

Hashimoto, H., Morikawa, H., Yamada, Y. & Kimura, A. (1985). A novel method for transformation of intact yeast cells by electroinjection of plasmid DNA. *Applied Microbiology and Biotechnology* **21**, 336-339.

Hebert, T. T. (1963), Precipitation of plant viruses by polyethylene glycol. *Phytopathology* **53**, 362.

Hibi, T. (1989). Electrotransfection of plant protoplasts with viral nucleic acids. *Advances in Virus Research* **3**, 329-342.

Hinnen, R., Hicks, J. B. & Fink, G. R. (1978). Transformation of yeast. *Proceedings of the National Academy of Sciences, USA* **75**, 1929-1933.

Jain, M. (1988). *Introduction to Biological Membranes*. John Wiley & Sons Inc.: New York.

Johnston, S. A., Anziano, P. Q., Shark, K., Sanford, J. C. & Butow, R. A. (1988). Mitochondrial transformation in yeast by bombardment with micro-projectiles. *Science* **240**, 1538-1541.

Karube, I., Tamiye, E. & Matsuoka, H. (1985). Transformation of *Saccharomyces cerevisiae* spheroplasts by high electric pulse. *FEBS Letters* **182**, 90-94.

Klebe, R. J., Harriss, J. V., Sharp, Z. J. & Douglas, M. G. (1983). A general method for polyethylene-glycol-induced genetic transformation of bacteria and yeast. *Gene* **5**, 333-341.

Klein, T. M., Wolf, E. D., Wu, R. & Sanford, J. C. (1987). High-velocity microprojectiles for delivering nucleic acids into living cells. *Nature* **327**, 70-73.

Klein, T. M., Harper, E. C., Svab, Z., Sanford, J. C., Fromm, M. E. & Maliga, P. (1988). Stable genetic transformation of intact *Nicotiana* cells by the particle bombardment process. *Proceedings of the National Academy of Sciences, USA* **85**, 8502-8505.

Klein, T. M., Kornstein, L., Sanford, J. C. & Fromm, M. E. (1989). Genetic transformation of maize cells by particle bombardment. *Plant Physiology* **91**, 440-444.

Lindsey, K. & Jones, M. G. K. (1987). The permeability of electroporated cells and protoplasts of sugar beet. *Planta* **172**, 346-355.

McCabe, D. E., Swain, W. F., Martinell, B. J. & Christou, P. (1988). Stable transformation of soybean (*Glycine max*) by particle acceleration. *Bio/Technology* **6**, 923-926.

McIntyre, D. A. & Harlander, S. K. (1989). Genetic transformation of intact *Lactococcus lactis* subsp. *lactis* by high-voltage electroporation. *Applied and Environmental Microbiology* **55**, 604-610.

MacNeil, D. I. (1987). Introduction of plasmid DNA into *Streptomyces lividans* by electroporation. *FEMS Microbiology Letters* **42**, 239-244.

Meilhoc, E., Masson, J. -M. & Teissié, J. (1990). High efficiency transformation of intact yeast cells by electric field pulses. *Bio/Technology* **8**, 223-227.

Miller, J. F., Dower, W. J. & Tompkins, L. S. (1988), High-voltage electroporation of bacteria: genetic transformation of *Campylobacter jejuni* with plasmid DNA. *Proceedings of the National Academy of Sciences, USA* **85**, 856-860.

Morikawa, H., Iida, A., Matsui, C., Ikegami, M. & Yamada, Y. (1986). Transfer into intact plant cells by electroinjection through cell walls and membranes. *Gene* **41**, 121-124.

Motoyoshi, F., Bancroft, I. B. & Watts, J. W. (1973). A direct estimate of the number of cowpea chlorotic mottle virus particles absorbed by tobacco protoplasts that become infected. *Journal of General Virology* **21**, 159-161.

Motoyoshi, F., Watts, J. W. & Bancroft, J. B. (1974). Factors influencing the infection of tobacco protoplasts by cowpea chlorotic mottle virus. *Journal of General Virology* **25**, 245-256.

Negrutiu, I., Shillito, R., Potrykus, I., Biasini, G. & Sala, F. (1987). Hybrid genes in the analysis of transformation conditions. I. Setting up a simple method for direct gene transfer in plant protoplasts. *Plant Molecular Biology* **8**, 363-373.

Neumann, E., Schaefer-Ridder, M., Wang, Y. & Hofschneider, P. H. (1982). Gene transfer into mouse lyoma cells by electroporation in high electric fields. *EMBO Journal* **1**, 841-845.

Nishiguchi, M., Sato, T. & Motoyoshi, F. (1987). An improved method for electroporation in plant protoplasts: infection of tobacco protoplasts by tobacco mosaic virus particles. *Plant Cell Reports* **6**, 90-93.

Okada, K., Nagata, T. & Takebe, I. (1986). Introduction of functional RNA into plant protoplasts by electroporation. *Plant Cell Physiology* **27**, 619-626.

Otsuki, Y., Takebe, I., Honda, Y. & Matsui, C. (1972). Ultrastructure of infection of tobacco mesophyll protoplasts by tobacco mosaic virus. *Virology* **49**, 188-194.

Potter, H. (1988). Electroporation in biology: methods, applications and instrumentation. *Analytical Biochemistry* **17**, 361-373.

Potter, H., Weir, L. & Leder, P. (1984). Enhancer-dependent expression of human κ immunoglobulin genes introduced into mouse pre-B lymphocytes by electroporation. *Proceedings of the National Academy of Sciences, USA* **81**, 7161-7165.

Sanford, J. C. (1988). The biolistic process. *Tibtec* **6**, 299-302.

Sanford, J. C., Klein, T. M., Wolf, E. D. & Allen, N. (1987). Delivery of substances into cells and tissues using a particle bombardment process. *Particulate Science and Technology* **5**, 27-37.

Scott, P. T. & Rood, J. I. (1989). Electroporation-mediated transformation of lysostaphin-treated *Clostridium perfringens*. *Gene* **82**, 327-333.

Shivarova, N., Forster, W., Jacob, H. -E. & Grigorova, R. (1983). Microbiological implications of electric field effects. VII. Stimulation of plasmid transformation of *Bacillus cereus* protoplasts by electric field pulses. *Zeitschrift für Allgemeine Mikrobiologie* **23**, 595-599.

Twell, D., Klein, T. & Fromm, M. E. & McCormick, S. (1989). Transient expression of chimeric genes delivered into pollen by microprojectile bombardment. *Plant Physiology* **91**, 1270-1274.

Varner, J. E. & Lin, L. -S. (1989). Plant cell wall architecture. *Cell* **56**, 231-239.

Watts, J. W., King, J. M. & Stacey, N. J. (1987). Inoculation of protoplasts with viruses by electroporation. *Virology* **157**, 40-46.

Zimmermann, U. & Vienken, J. (1982). Electric field-induced cell-to-cell fusion. *Journal of Membrane Biology* **6**, 165-182.

Chapter 4

Saccharomyces cerevisiae: a host for the production of foreign proteins

Jill E. Ogden

The yeast *Saccharomyces cerevisiae* has been developed over a number of years as a host for the production of foreign proteins. There are several features of yeast which make it an attractive option as a host expression system. First, yeast has a long history of use in the brewing and baking industries and is therefore accepted as safe for use in fermentation processes. Second, *S. cerevisiae* is amenable to genetic and recombinant DNA manipulation and there is much information on its metabolism and fermentation characteristics. Third, yeast, in contrast to Gram-negative bacteria such as *Escherichia coli*, is free of lipopolysaccharide (LPS) and associated problems of pyrogenicity.

There are very many examples of the production of a wide range of foreign proteins in *S. cerevisiae*, including mammalian, plant, fungal and bacterial proteins. These are too numerous to mention here. Instead, I hope that by providing a number of specific examples, to highlight the advantages and also the drawbacks of using *S. cerevisiae* as an expression system.

The basis of yeast expression systems

Ideally for production of a heterologous protein in yeast, the optimum expression cassette comprises efficient promoter sequences, which may be inducible or constitutive, the cDNA for the protein of interest and transcription termination sequences. Expression cassettes can be maintained on plasmids containing replication functions of the indigenous 2 μm plasmid or other autonomously replicating sequences, or they can be integrated into the yeast genome. The use of cDNAs is necessary as *S. cerevisiae* is unable to recognise and process higher eukaryotic introns (Beggs *et al.*, 1980). It is also important to consider the nature of the protein to be produced as this may determine whether intracellular expression or secretion of the product is preferable.

Promoters

Yeast promoters have been studied extensively over the last few years and this has revealed a number of key control elements which are required for optimal expression and accurate mRNA initiation. This work is beyond the scope of this chapter and has been reviewed in detail elsewhere (Struhl, 1986; Guarente, 1988). One of the control elements identified, the Upstream Activation Sequence (UAS), is normally located several hundred nucleotides upstream from the initiation methionine of the coding region, and is required for maximal gene expression. In many gene promoters the UAS is the binding site for regulatory proteins. This has been exploited for use in heterologous expression studies by the construction of hybrid promoters which comprise the transcription regulatory regions of one promoter and the transcription initiation sequences of another (see later).

Many of the promoters used to date to direct heterologous gene expression are based on those of genes encoding glycolytic enzymes, such as phosphoglyceratekinase (*PGK*) (Mellor *et al.*, 1983), alcohol dehydrogenase 1 (*ADH1*) (Hitzeman *et al.*, 1981) and glyceraldehyde 3-phosphate dehydrogenase (*GAPDH*) (Bitter & Egan, 1984). These single copy genes encode some of the most abundant mRNA and protein species within the cell and therefore their promoters are obvious choices for directing foreign gene expression.

An alternative approach for obtaining promoters has been described by several groups (Goodey *et al.*, 1986; Santangelo *et al.*, 1988; Sleep *et al.*, 1991a). This entails the selection of promoter sequences from a gene library based on their efficiency in directing expression of a heterologous gene which imparts a particular selective advantage on the yeast cell. This method has proved useful in isolating some very effective promoter systems.

Yeast promoters can be loosely classed into those that confer constitutive expression and those that are regulated. The glycolytic gene promoters are on the whole constitutive, although there is evidence that the *PGK* promoter is subject to some degree of regulation depending on carbon source (Stanway *et al.*, 1987). Regulatable promoters are often regarded as preferable for heterologous gene expression. Thus yeast cultures can be grown to high biomass before they are required to synthesise the protein of interest. Obviously this is important for production of proteins which are potentially toxic to the yeast cell. The genes involved in galactose metabolism, such as the *GAL1* (galactokinase), *GAL10* (uridine diphosphogalactose-4-epimerase) and *GAL7* (α-D-galactose-1-phosphate uridyltransferase) genes, offer classical examples of the mechanism of regulation in yeast (for review, see Johnston, 1987). These three genes are clustered on chromosome II where *GAL1* is transcribed divergently from the *GAL10* and *GAL7* genes (St. John & Davis, 1981). The promoters of these genes contain homologous upstream activation sequences (UAS_{GAL}) which, in the presence of galactose, are binding sites for the positive regulatory protein produced by the *GAL4* gene resulting in an approximate 1000-fold induction in gene expression (Giniger, Varnum & Ptashne, 1985). In the absence of galactose the action of the GAL4 protein is inhibited by direct interaction with the specific negative regulator produced by the *GAL80* gene (Lue *et al.*, 1987; Ma & Ptashne, 1987). The galactose-inducible genes are also subject to glucose or catabolite repression (St. John & Davis, 1981).

Many groups have described the expression of heterologous proteins using the regulatable *GAL1*, *GAL10* and *GAL7* promoters. Expression of heterologous products appears to be fairly tightly regulated by the galactose/glucose system. For example, studies of the expression of calf prochymosin (Goff *et al.*, 1984) using the *GAL1* promoter demonstrated that no foreign protein was detected when the yeast cells were grown in glucose media. However, in many cases the levels of heterologous protein produced after galactose induction have been somewhat disappointing. It is believed that this results from the limited amount of GAL4 protein available to induce transcription from both heterologous UASGAL-containing expression cassettes on multicopy plasmids (see later) and the indigenous galactose-inducible genes (Baker *et al.*, 1987). Integration of extra copies of the *GAL4* gene into the genome has improved expression levels; however, because of titration of GAL80 regulator by increased levels of GAL4 protein, galactose regulation of expression can be compromised (Johnston & Hopper, 1982). Schultz *et al.* (1987) describe an alternative approach to this by integrating copies of a *GAL4* gene that is itself under the control of the galactose-regulated *GAL10* promoter. Thus, in the absence of galactose only basal levels of GAL4 protein are produced from the 'wild type' constitutive *GAL4* gene; on induction with galactose, the hybrid *GAL4* gene also contributes to the production of GAL4 protein.

A number of other regulatable yeast promoters have been employed for heterologous protein production; e.g. the promoter of the repressible acid phosphatase gene *PHO5*, which responds to depletion of inorganic phosphate in the culture medium (Lemire *et al.*, 1985; Nakao *et al.*, 1986). The *PHO5* promoter has been used successfully to direct expression of several proteins (Kramer *et al.*, 1984; Izumoto *et al.*, 1987). However, removal of inorganic phosphate from the medium is not regarded as an efficient or practical method of induction on a large scale.

More preferable is the use of promoters of genes whose regulation is dependent on the physiological status of the cell, such as the changes correlated with the utilisation of glucose. Such promoters are subject to catabolite repression during the initial phases of growth on glucose, when biomass accumulates, and are subsequently released from repression

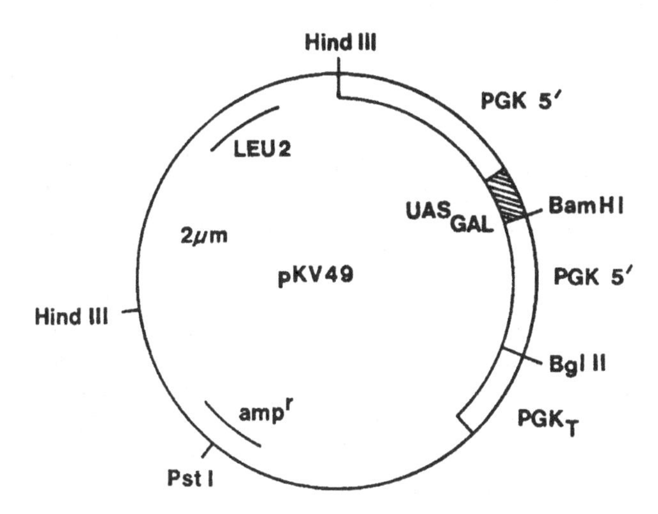

Fig. 4.1. A typical PAL promoter expression vector. PGK 5′ and PGK_T are promoter and terminator sequences from the phosphoglycerate kinase gene; UAS_{GAL} is the upstream activation sequence of the *GAL1-10* genes. Heterologous cDNAs are inserted into the unique *Bgl*II expression site.

once the yeast cell switches to respiratory metabolism and utilises the available ethanol. Promoters of this type include those of the iso-1-cytochrome c (*CYC1*) (Smith *et al.*, 1979), alcohol dehydrogenase II (*ADHII*) (Denis, Ciriacy & Young, 1981) and protease B (*PRB1*) (Moehle *et al.*, 1987) genes. The *PRB1* promoter has proved very efficient for the production of human serum albumin (Sleep *et al.*, 1991b).

In an approach to producing alternative regulatable systems, many groups describe what could be called 'designer' promoters, which combine the regulatory regions of one promoter with the transcription initiation sequences of another. One example of the hybrid promoter is the PAL promoter, in which the native UAS of the *PGK* promoter has been replaced by the UAS_{GAL} (Cousens, Wilson & Hinchliffe, 1990). Expression from the PAL promoter is subject to glucose repression but induced when the cells are grown on galactose as carbon source. A typical PAL expression vector is shown in Fig. 4.1. Further examples of hybrid promoters include *GAPDH*-UAS_{GAL} (Bitter & Egan, 1988) and

GAPDH-ADHII (Cousens *et al.*, 1987), which are regulated by galactose and glucose, respectively.

An alternative version of the hybrid promoter is a bi-directional expression system, which contains two different expression cassettes under the control of the divergent UAS_{GAL} from the *GAL1-GAL10* intergenic region (Johnston & Davis, 1984). This system allows co-expression of two different proteins within the cell and also offers an approach to the production of subunit proteins in yeast (Kenny & Hinchliffe, 1989).

Other methods for co-expressing proteins have been described. Independent expression cassettes can be introduced into yeast on the same plasmid (Horowitz *et al.*, 1990) or by co-transformation of two separate plasmids carrying complementary selectable markers. The co-transformation approach has been used successfully by Wood *et al.* (1985), who demonstrated that functional monoclonal antibodies could be assembled in yeast by co-expressing their component heavy and light chains.

Terminators

To terminate transcription from heterologous cDNAs, yeast expression vectors contain DNA fragments of several hundred base pairs in length, which contain the 3′ untranslated (terminator) regions of yeast genes such as *PGK*, *CYC1* and *ADH1*. Efficient termination of mRNA transcripts is essential for optimal gene expression. In *S. cerevisiae* the processes involved in transcription termination and subsequent mRNA 3′ end formation are less well characterised than in higher eukaryotes, where the conserved sequence AATAAA is essential for correct processing (reviewed by Birnstiel, Busslinger & Struhl, 1985). Analogous to the AATAAA element in higher eukaryotes, the consensus sequences TAG...TAGT or TATGT...(AT rich)...TTT have been proposed as a key element in transcription termination for many yeast genes, including *CYC1* (Zaret & Sherman, 1982). More recently, Osborne & Guarente (1989) have suggested that rather than a particular sequence element signalling 3′ end formation, transcription termination is determined more generally by sequences high in AT content. This can be a potential problem for expression of heterologous proteins whose cDNAs are AT rich. For example, a cDNA encoding fragment C, a 50 kD polypeptide derived from tetanus toxin which is approximately 75% AT rich, failed to be expressed in *S. cerevisiae* because of fortuitous transcription termination events in several regions of the DNA sequence (Romanos *et al.*, 1990). Efficient transcription and protein production was achieved only after elimination of the terminators by synthesising the cDNA with an increased GC content.

Stability of the expression cassette

The majority of vector systems used in *S. cerevisiae* are based on the indigenous 2μm plasmid, which is present in most strains (for review see Broach, 1983; Futcher, 1988; Knowlton, 1986). 2μm-based plasmids are maintained at high copy numbers (20 to 200 copies per haploid cell) and are relatively stable. In addition to the sequences required for heterologous protein production, yeast expression vectors contain an origin of replication and selectable markers for propagation in *E. coli*, most commonly derived from pBR322, and the yeast 2μm origin of replication *STB* or *REP3* and a yeast selectable marker. The 2μm trans-acting products of the *REP1*, *REP2* and *FLP* genes, which are required for plasmid maintenance and copy number control (Jayaram, Li & Broach, 1983), are provided by indigenous 2μm plasmids in [cir^+] strains.

Yeast selectable markers are often genes which complement auxotrophic mutations such as *LEU2*, *TRP1* and *URA3*. Two different 2μm-based plasmids use *LEU2* as selectable marker: YEp13 (Broach, 1983) and pJDB207 (Beggs, 1978). These are essentially the same except that the *LEU2* gene present on pJDB207-based plasmids is defective in that it lacks a functional promoter, and is consequently known as *LEU2d*. It is believed that the poor expression of *LEU2d* results in higher plasmid copy numbers than the equivalent *LEU2* selectable plasmids (Erhart & Hollenberg, 1983).

Selectable markers for yeast are not confined to those complementing auxotrophic mutations. The *CUP1* gene which confers copper resistance has been used successfully, specifically for yeasts which are not easily manipulated genetically, such as brewing yeasts (Henderson, Cox & Tubb, 1985). For more detail on dominant selectable markers see Chapter 8.

For the large scale production of foreign protein, yeast must be grown over many generations (typically 40 to 50) and, even under selective growth conditions, this results in plasmid loss at approximately 1-2% per generation (Murray & Szostak, 1983). One possible cause of such plasmid instability is intermolecular recombination between homologous sequences on the expression vector and indigenous 2μm plasmids. To overcome the inherent instability of conventional 2μm-based vector systems, Chinery & Hinchliffe (1989) have designed a vector which is as near as possible to authentic 2μm with inserted selectable marker and expression cassette (Fig. 4.2). One important feature of this vector system is the placement of the bacterial sequences, necessary for propagation in *E. coli*, within the 2μm inverted repeats at the *XbaI* restriction site. On transformation of such plasmids into [cir^0] yeast strains, recombination mediated by the FLP protein occurs between the inverted repeats at the

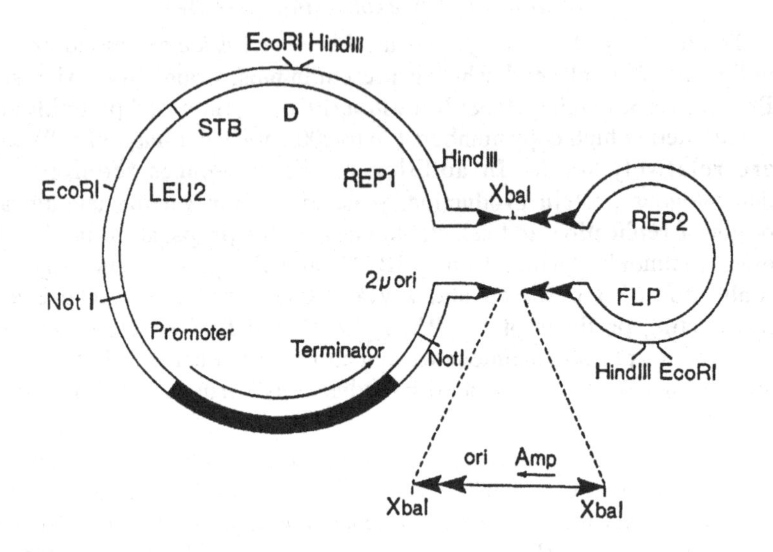

Fig. 4.2. A typical 'disintegration vector'. Bacterial sequences required for replication (*ori*) and selection (*amp*^{R}) in *E. coli* are flanked by two 74 bp repeat sequences (< <–>) containing the FLP protein recognition site. The expression cassette containing a promoter, cDNA encoding a heterologous protein (shaded box) and terminator was inserted as a *Not*I fragment into a unique *Not*I restriction site in the vector.

FLP recognition target site (*FRT*) resulting in the excision and subsequent loss of bacterial sequences. This is advantageous from the regulatory point of view. These so-called 'disintegration vectors' have high copy numbers (20 to 30) and have proved to be extremely stable over hundreds of generations of continuous culture (Collins, 1990; S. H. Collins, D. J. Mead & P. D. Moir, unpublished).

Other methods of introducing expression cassettes into yeast have been described. One of these systems relies on origins of replication derived from yeast chromosomal DNA, known as autonomous replication sequences or *ARS*. Vectors of this type, such as the *TRP* selectable plasmid YRp7 (Struhl *et al.*, 1979), have high copy numbers but are very unstable even under selective growth conditions. The introduction of DNA sequences containing the centromere regions (*CEN*) of several different chromosomes effectively stabilises *ARS*-based plasmids by making them into mini-chromosomes (Clarke & Carbon, 1980). However, the copy number of *ARS-CEN* vectors is reduced to approximately one per cell.

Mitotic stability of expression cassettes can be achieved by integrating the required DNA sequences into the yeast genome. However, these vectors have much lower copy numbers than the 2 μm-based plasmid vectors, even though multiple integrations do sometimes occur (Orr-Weaver & Szostak, 1983). Recently, Lopes *et al.* (1989) have described a vector which integrates into the genome at high copy numbers (100). This is achieved by targeting integration into the ribosomal DNA locus which comprises 100 to 200 tandemly repeated units (Petes, 1979). The resulting integrants appear to retain a high proportion of their plasmid copies over many generations of non-selective growth.

The site of accumulation of foreign proteins

It is important to consider the nature of a foreign protein before deciding whether to produce it intracellularly or by secretion into the culture medium. For example, it may not be possible for proteins that are normally located intracellularly to be translocated successfully through the secretory pathway because their structure precludes passage across biological membranes. In contrast, when normally secreted proteins are expressed intracellularly, some, but by no means all, form insoluble aggregates because of the absence of disulphide bond formation and correct folding (Moir *et al.*, 1985). Such insoluble heterologous products have been solubilised and refolded successfully, one example being human albumin (HA) (Burton, Quirk & Wood, 1989). However, this approach is both costly and time consuming and is, in many cases, prohibitive for large scale production.

Secretion

Given these considerations, secretion of foreign proteins from *S. cerevisiae* has often been the preferred route. The yeast secretory pathway shares many features with that of higher eukaryotes and has been reviewed extensively elsewhere (Schekman & Novick, 1982). One advantage of the secretion approach is that *S. cerevisiae* does not normally secrete many proteins into the culture medium, thus facilitating subsequent purification of the heterologous product. For secretion, proteins require a hydrophobic N-terminal 'leader' or 'signal' sequence of around 20 amino acids, which targets the newly translated polypeptide through the secretory pathway. Maturation of the N-terminus of the polypeptide depends on a number of sequential enzyme steps in the endoplasmic reticulum (ER) and Golgi apparatus, which cleave specific sites in the leader sequence before the protein is exported to the outside of the cell. Native yeast leader sequences, such as those of the α mating pheromone (*MFα1*) (Kurjan & Herskowitz, 1982), acid phosphatase (*PHO5*) (Arima *et al.*, 1983) and invertase (*SUC2*) (Carlson *et al.*, 1983) genes, have been used successfully for secretion of foreign proteins.

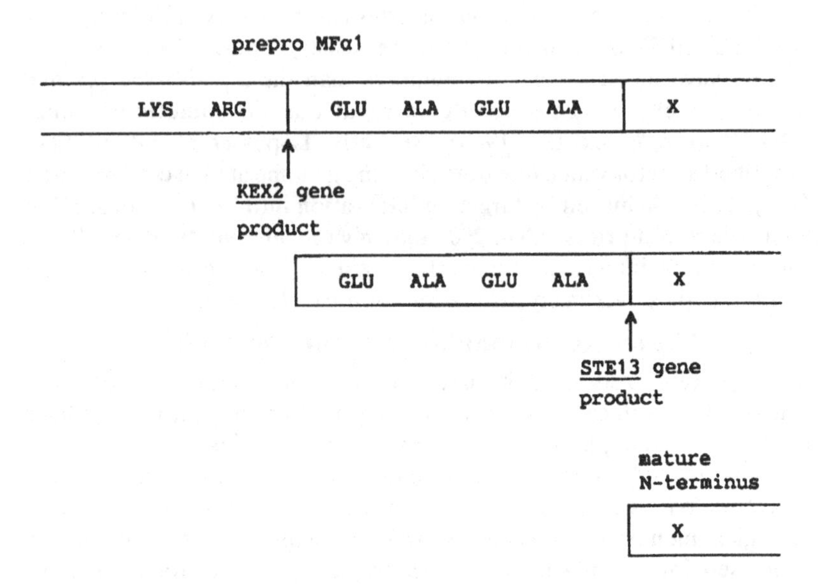

Fig. 4.3. Schematic diagram showing steps in the processing of the MFα1 prepro leader sequence to release the authentic N-terminus of a secreted heterologous protein (X).

Probably the most widely used leader sequence is that of the *MFα1* gene. This consists of a pre-region of 19 amino acids, which resembles other classical signal sequences, followed by a much longer pro-region of 70 amino acids (Kurjan & Herskowitz, 1982). The MFα1 pro-sequence is processed sequentially by the endopeptidase products of the *KEX2* and *STE13* genes. The KEX2 endopeptidase cleaves specifically after a pair of basic amino acids (lys, arg) and the STE13 dipeptidyl endopeptidase removes a tetrapeptide (glu, ala) (Julius *et al.*, 1983, 1984) (Fig. 4.3). Some authors have reported that use of a complete α-factor leader results in secreted products with heterogeneous N-termini due to incomplete processing by the STE13 endopeptidase (Bitter *et al.*, 1984). Such observations are consistent with the STE13 protease being rate-limiting. One method of overcoming this problem has been to use a modified prepro α-factor leader which encodes the KEX2 lys-arg cleavage site but lacks the glu-ala repeats. The use of such a leader, where correct N-termini processing is independent of the STE13 protease step, has proved successful in improving efficiency of secretion (Brake *et al.*, 1984).

Some heterologous proteins have been secreted from yeast using their own or related mammalian leader sequences. Examples include the human proteins antithrombin III (Bröker, Ragg & Karges, 1987), lysozyme (Jigami *et al.*, 1986) and albumin (Sleep, Belfield & Goodey, 1990), wheat α-amylase (Rothstein *et al.*, 1987), *Aspergillus awamori* glucoamylase (Innis *et al.*, 1985) and *Trichoderma reesei* cellobiohydrolases (Penttilä *et al.*, 1988). Where analyzed, N-terminal processing of these products was authentic indicating common features among eukaryotic processing signals. Indeed, studies on the maturation of human serum albumin (HSA) demonstrate that processing of the native leader sequence requires an endopeptidase which is functionally equivalent to the yeast KEX2 enzyme, and which cleaves the leader after a pair of basic residues (arg, arg) (Bathurst *et al.*, 1987). Sleep *et al.* (1990a) compared the efficiency of five KEX2-type leaders, including prepro MFα1, native HSA and an HSA/prepro MFα1 fusion leader, for secretion of human albumin from yeast. Their studies demonstrate that the secretion leader can affect not only levels of secreted protein but also the quality of the product.

In addition to maturation of N-terminal sequences, post-translational processing of secreted proteins in eukaryotes includes the addition of carbohydrate to specific amino acid residues. Glycosylation in yeast is of both the N-linked (*via* an asparagine amide) and O-linked (*via* a serine or threonine hydroxyl) types, occurring at the sequences Asn-X-Ser/Thr and Thr/Ser respectively (see Tanner & Lehle, 1987 for review). Core glycosylation events occur in the endoplasmic reticulum and completion of the glycosylation process occurs in the Golgi apparatus (Esmon, Novick & Schekman, 1981). Glycosylation in *S. cerevisiae* is of the high mannose type, with mannose chains of between 50 and 150 residues added to the core carbohydrate. For heterologous proteins this can result in expression of 'overglycosylated' products of heterogeneous size, containing a much higher proportion of carbohydrate than in their native form. This has been shown for a range of proteins including erythropoietin (Elliott *et al.*, 1989), α_1-antitrypsin (Moir & Dumais, 1987) and urokinase-type plasminogen activator (u-PA) (Melnick *et al.*, 1990; J. Steven, unpublished). Moreover, frequently a large proportion of the protein fails to be secreted into the culture medium and remains cell associated, either in the periplasmic space or somewhere within the secretory pathway (Smith, Duncan & Moir, 1985).

One approach to overcoming the problem of hyperglycosylation of yeast-derived glycoproteins is the removal of the glycosylation site. This has proved effective for u-PA, where the single N-linked site was abolished by changing asparagine 302 to an alanine using site-directed mutagenesis (Melnick *et al.*, 1990). The resulting product lacked any detectable N-linked glycosylation, however it is possible that the absence of carbohydrate may affect the stability of the protein (J. Steven, unpublished).

A further approach to reducing the high mannose content of *S. cerevisiae* glycoproteins is the use of strains carrying mutations which affect mannose chain elongation, namely *mnn9* and *pmr1*, also known as *ssc1*. Strains carrying the *mnn9* mutation, first identified by Ballou and coworkers, are defective in the addition of mannose residues to the ER core oligosaccharide (Tsai, Frevert & Ballou, 1984). The α_1-antitrypsin secreted from an *mnn9* mutant strain was found to be homogeneous and of similar size to the plasma-derived protein (Moir & Dumais, 1987), consistent with only core glycosylation occurring at the three N-linked sites on the protein.

The *PMR1* gene codes for a Ca^{2+} ATPase and is identical to *SSC1* (Rudolph *et al*., 1989). A mutation of the *SSC1* gene (*ssc1-1*) was originally identified on the basis of a 'supersecretion' phenotype for the production of calf prochymosin (Smith *et al*., 1985). Subsequent studies demonstrated that the secretion of bovine growth hormone (Smith *et al*., 1985) and u-PA (Melnick *et al*., 1990) was also more efficient in *ssc1-1* compared with wild type strains. That the *ssc1-1* mutation affects glycosylation has been demonstrated using yeast invertase, α_1-antitrypsin and u-PA (Rudolph *et al*., 1989). As with *mnn9* strains the addition of mannose residues to the core oligosaccharide appears to be blocked. Taken together, the increased secretion efficiency and absence of outer mannose chains in *pmr1/ssc1-1* strains suggests a bypass of some steps in the secretory pathway, most likely the later steps involving the Golgi apparatus.

Another potential drawback of *S. cerevisiae* for secretion of mammalian glycoproteins is its inability to carry out the more complex carbohydrate modifications of higher eukaryotes, such as the addition of galactose and sialic acid residues. Thus, certain proteins secreted by yeast may be very different from their native form. Although many such yeast-derived proteins are biologically active, the effect of differences in carbohydrate on the immunogenicity, half life and physiological targeting of proteins for therapeutic use remains to be seen. Of course, for normally non-glycosylated proteins this is not a problem.

Intracellular production

As mentioned earlier, one potential drawback of intracellular production of heterologous proteins in *S. cerevisiae* is the possibility of formation of insoluble inclusion bodies. This is not always confined to proteins which require disulphide bond formation. However, there are many reports of the successful expression of soluble biologically active heterologous products in yeast, and some of these display very high levels of expression. There are advantages to intracellular expression. For example, it is not unreasonable to assume that much higher levels can be obtained for some proteins using the intracellular route, rather than secretion, because of potential rate-limiting steps in the secretory pathway. Furthermore, proteins produced cytoplasmically are not glycosylated and this avoids the problems of overglycosylation that can arise.

For many cytosolic proteins, translation initiation at the methionine codon is followed by co-translational removal of the methionine residue by methionylaminopeptidase (MAP). The specificity of MAP is governed by the identity of the amino acid residue immediately following the initiation methionine (Huang *et al.*, 1987). Varshavsky and co-workers have recently described what they term the N-terminal rule, in which specific N-terminal amino acids are ranked according to the degree of metabolic stability they confer on proteins (Bachmair, Finley & Varshavsky, 1986). For instance, methionine is very stable whilst arginine is not. In general, those amino acids which destabilise proteins also inhibit the removal of the N-terminal methionine by MAP. These 'destabilising' amino acids are often found at the processed N-terminus of secreted proteins. Thus cytosolic expression of many naturally secreted eukaryotic proteins can result in retention of the extra methionine residue at the N-terminus. This is undesirable for proteins to be used as therapeutics because of potential immunogenicity problems. One method of overcoming this is to fuse the yeast ubiquitin gene upstream of the heterologous cDNA (Sabin *et al.*, 1989). Following expression, the fusion protein is cleaved specifically by an endogenous ubiquitin-specific endopeptidase, to release the authentic N-terminus of the protein. In addition, the use of ubiquitin fusions has been shown to improve dramatically the expression levels of a number of heterologous products (Ecker *et al.*, 1989).

Although it is generally regarded as undesirable to produce heterologous products as intracellular, insoluble aggregates, there are notable exceptions to this. In the early 1980s Valenzuela, Medina & Rutter (1982) demonstrated that *S. cerevisiae* could be used to produce hepatitis B virus surface antigen (HbsAg) as 20 nm virus-like particles, similar to those found in human serum. The particles purified from yeast were immunogenic. Subsequent development has resulted in the production of a vaccine which is safe in humans (Jilg *et al.*, 1984; Scolnick *et al.*, 1984) and is effective in protecting chimpanzees from hepatitis B (McAleer *et al.*, 1984). This recombinant vaccine has now been approved for clinical use.

A similar approach to the production of antigens using yeast utilises the virus-like particles (VLPs) produced by the indigenous retrotransposon, Ty (Mellor *et al.*, 1985; Mellor, Kingsman & Kingsman, 1986, for review). Construction of gene fusions containing the portion of Ty required for VLP formation and various foreign genes results in the production of hybrid VLPs (Adams *et al.*, 1987). Hybrid particles containing parts of Human Immunodeficiency Virus (HIV) gp120 coat protein and an Influenza virus Haemagglutinin (HA) have been produced. Ty:HIV and Ty:HA VLPs raise an immunogenic response in rabbits against the non Ty components. Hybrid Ty-VLPs are now under development as a source of novel vaccines.

Improving yields

If yeast is to be used in a commercially viable process for producing a heterologous protein, the yield of the product is of paramount importance. Methods for improving yields have often relied on assessing a large number of different host yeast strains. There can be quite striking differences between the expression levels of a given protein using the same vector system in different yeasts. This is presumably a reflection of the varying genetic background of the yeast strains. The genetic manipulation of appropriate yeasts can result in progeny with the combined favourable characteristics. Yeast strains defective in protease synthesis, e.g. those carrying the pleiotropic *pep4-3* mutation which results in a deficiency of several proteases (Jones, 1984), have been shown to produce better quality heterologous proteins (Cabezón *et al.*, 1984; Gardell *et al.*, 1990).

Mutation screens for enhanced production of foreign proteins have been described by several groups. For secreted proteins screening for up-mutants relies on enzymatic or immuno-assays, where the secreted product is detected as a halo round the yeast colony (Smith *et al.*, 1985; Sakai, Shimizu & Hishinuma, 1988; Suzuki, Ichikawa & Jigami, 1989; Sleep *et al.*, 1991b). Using this type of approach improvements in secretion yields of up to 18 fold have been reported. Curing the yeast strains of plasmid followed by retransformation demonstrates that mutations carried by the host yeast genome are responsible for the overproducing phenotype. In most cases the precise mechanisms of the mutant phenotypes have yet to be elucidated.

Mutation screening for enhanced production of intracellular proteins is not as straightforward as described for secreted proteins. However, in many instances yeasts selected for enhanced production of a secreted product have also been shown to produce high levels of intracellular proteins. This is exemplified by the observation that mutant strains selected for their ability to secrete enhanced levels of human albumin also produce very high levels of human plasminogen activator inhibitor type 2 (PAI-2) and α_1-antitrypsin Pittsburgh variant (α_1AT-P), at levels of 20% and 40% total soluble protein, respectively (Sleep *et al.*, 1991b).

The achievement of such yields in combination with the ability to grow yeast to high biomass on a large scale offers an ideal system for the production of proteins economically. As discussed in this chapter *S. cerevisiae* may not be a suitable host for certain therapeutic proteins and therefore it is important to consider every protein, case by case. However, as our knowledge of the molecular genetics and physiology of yeast increases it is possible that at least some of the drawbacks may be overcome.

Acknowledgements I would like to thank my colleagues Drs Steven Collins, David Mead, Peter Moir and John Steven for allowing me to use their unpublished data and Dr Michael Courtney for critical reading of the manuscript. I thank Joanne Middleton for typing the manuscript.

References

Adams, S. E., Dawson, K. M., Gull, K., Kingsman, S. M. & Kingsman, A. J. (1987). The expression of hybrid HIV:Ty virus-like particles in yeast. *Nature* **329**, 68-70.

Arima, K., Oshima, T., Kubota, I., Nakamura, N., Mizunaga, T. & Toh-e, A. (1983). The nucleotide sequence of the yeast *PHO5* gene: a putative precursor of repressible acid phosphatase contains a signal peptide. *Nucleic Acids Research* **11**, 1657-1672.

Bachmair, A., Finley, D. & Varshavsky, A. (1986). *In vivo* half-life of a protein is a function of its amino-terminal residue. *Science* **234**, 179-186.

Baker, S. M., Johnston, S. A., Hooper, J. E. & Jaehning, J. A. (1987). Transcription of multiple copies of the yeast *GAL7* gene is limited by specific factors in addition to *GAL4*. *Molecular and General Genetics* **208**, 127-134.

Bathurst, I. C., Brennan, S. O., Carrell, R. W., Cousens, L. S., Brake, A. J. & Barr, P. J. (1987). Yeast KEX2 protease has the properties of a human proalbumin converting enzyme. *Science* **235**, 348-357.

Beggs, J. D. (1978). Transformation of yeast by a replicating hybrid plasmid. *Nature* **275**, 104-109.

Beggs, J. D., van den Berg, J., van Ooyen, A. & Weissmann, C. (1980). Abnormal expression of chromosomal rabbit β-globin gene in *Saccharomyces cerevisiae*. *Nature* **283**, 835-840.

Birnstiel, M. L., Busslinger, M. & Struhl, K. (1985). Transcription termination and 3′ processing: the end is in site. *Cell* **41**, 349-359.

Bitter, G. A. & Egan, K. M. (1984). Expression of heterologous genes in *Saccharomyces cerevisiae* from vectors utilising the glyceraldehyde-3-phosphate dehydrogenase gene promoter. *Gene* **32**, 263-274.

Bitter, G. A. & Egan, K. M. (1988). Expression of interferon-gamma from hybrid yeast *GPD* promoters containing upstream regulatory sequences from the *GAL1-GAL10* intergenic region. *Gene* **69**, 193-207.

Bitter, G. A., Chen, K. K., Banks, A. R. & Lai, P. -H. (1984). Secretion of foreign proteins from *Saccharomyces cerevisiae* directed by α-factor gene fusions. *Proceedings of the National Academy of Sciences, USA* **81**, 5330-5334.

Brake, A. J., Merryweather, J. P., Coit, D. G., Heberlein, U. A., Masiarz, F. R., Mullenbach, G. T., Urdea, M. S., Valenzuela, P. & Barr, P. J. (1984).

α-factor-directed synthesis and secretion of mature foreign proteins in *Saccharomyces cerevisiae*. *Proceedings of the National Academy of Sciences, USA* **81**, 4642-4646.

Broach, J. R. (1983). Construction of high copy yeast vectors using 2 μm circle sequences. *Methods in Enzymology* **101**, 307-325.

Bröker, M., Ragg, H. & Karges, H. E. (1987). Expression of human antithrombin III in *Saccharomyces cerevisiae* and *Schizosaccharomyces pombe*. *Biochimica et Biophysica Acta* **908**, 203-213.

Burton, S. J., Quirk, A. V. & Wood, P. C. (1989). Refolding human serum albumin at relatively high protein concentration. *European Journal of Biochemistry* **179**, 379-387.

Cabezón, T., De Wilde, M., Herion, P., Loriau, R. & Bollen, A. (1984). Expression of human α1-antitrypsin cDNA in the yeast *Saccharomyces cerevisiae*. *Proceedings of the National Academy of Sciences, USA* **81**, 6594-6598.

Carlson, M., Taussig, R., Kustri, S. & Botstein, D. (1983). The secreted form of invertase in *Saccharomyces cerevisiae* is synthesised from mRNA encoding a signal sequence. *Molecular and Cellular Biology* **3**, 409-447.

Chinery, S. A. & Hinchliffe, E. (1989). A novel class of vector for yeast transformation. *Current Genetics* **16**, 21-25.

Clarke, L. & Carbon, J. (1980). Isolation of a yeast centromere and construction of functional small circular chromosomes. *Nature* **287**, 504-509.

Collins, S. H. (1990). Production of secreted proteins in yeast. In *Protein Production by Biotechnology*, (ed. T. J. R. Harris), pp. 61-77. Elsevier Science Publishers: New York.

Cousens, D. J., Wilson, M. J. & Hinchliffe, E. (1990). Construction of a regulated *PGK* expression vector. *Nucleic Acids Research* **18**, 1308.

Cousens, L. S., Shuster, J. R., Gallegos, C., Ku, L. L., Stempien, M. M., Urdea, M. S., Sanchez-Pescador, R., Taylor, A. & Tekamp-Olson, P. (1987). High level expression of proinsulin in the yeast, *Saccharomyces cerevisiae*. *Gene* **61**, 265-275.

Denis, C. L., Ciriacy, M. & Young, E. T. (1981). A positive regulatory gene is required for accumulation of the functional mRNA for the glucose-repressible alcohol dehydrogenase from *Saccharomyces cerevisiae*. *Journal of Molecular Biology* **148**, 355-368.

Ecker, D. J., Stadel, J. M., Butt, T. R., Marsh, J. A., Monia, B. P., Powers, D. A., Gorman, J. A., Clark, P. E., Warren, F., Shatzman, A. & Crooke, S. T. (1989). Increasing gene expression in yeast by fusion to ubiquitin. *Journal of Biological Chemistry* **264**, 7715-7719.

Elliott, S., Giffin, J., Suggs, S., Lau, E. P. & Banks, A. R. (1989). Secretion of glycosylated human erythropoietin from yeast directed by the α-factor leader region. *Gene* **79**, 167-180.

Erhart, E. & Hollenberg, C. P. (1983). The presence of a defective *LEU2* gene on 2 μ DNA recombinant plasmids of *Saccharomyces cerevisiae* is responsible for curing and high copy number. *Journal of Bacteriology* **156**, 625-635.

Esmon, B., Novick, P. & Schekman, R. (1981). Compartmentalized assembly of oligosaccharides on exported glycoproteins in yeast. *Cell* **25**, 451-460.

Futcher, A. B. (1988). The 2 μm circle plasmid of *Saccharomyces cerevisiae*. *Yeast* **4**, 27-40.

Gardell, S. J., Hare, T. R., Han, J. H., Markus, H. Z., Keech, B. J., Carty, C. E., Ellis, R. W. & Schultz, L. D. (1990). Purification and characterisation of human plasminogen activator inhibitor type 1 expressed in *Saccharomyces cerevisiae*. *Archives of Biochemistry and Biophysics* **278**, 467-474.

Giniger, E., Varnum, S. & Ptashne, M. (1985). Specific DNA binding of GAL4, a positive regulatory protein of yeast. *Cell* **40**, 767-774.

Goff, C. G., Moir, D. T., Kohno, T., Gravius, T. C., Smith, R. A., Yamasaki, E. & Taunton-Rigby, A. (1984). Expression of calf prochymosin in *Saccharomyces cerevisiae*. *Gene* **27**, 35-46.

Goodey, A. R., Doel, S. M., Piggott, J. R., Watson, M. E. E., Zealey, G. R., Cafferkey, R. & Carter, B. L. A. (1986). The selection of promoters for the expression of heterologous genes in the yeast *Saccharomyces cerevisiae*. *Molecular and General Genetics* **204**, 505-511.

Guarente, L. (1988). UASs and enhancers: common mechanism of transcriptional activation in yeast and mammals. *Cell* **52**, 303-305.

Henderson, R. C. A., Cox, B. S. & Tubb, R. S. (1985). The transformation of brewing yeast with a plasmid containing the gene for copper resistance. *Current Genetics* **9**, 133-136.

Hitzeman, R. A., Hagie, F. E., Levine, H. L., Goeddel, D. V., Ammerer, G & Hall, B. D. (1981). Expression of a human gene for interferon in yeast. *Nature* **293**, 717-722.

Horowitz, B., Eakle, K. A., Scheiner-Bobis, G., Randolph, G. R., Chen, C. Y., Hitzeman, R. A. & Farley, R. A. (1990). Synthesis and assembly of functional mammalian Na,K-ATPase in yeast. *Journal of Biological Chemistry* **265**, 4189-4192.

Huang, S., Elliott, R. C., Liu, P. -S., Koduri, K., Weickmann, J. L., Lee, J. -H., Blair, L. C., Ghosh-Dastidar, P., Bradshaw, R. A., Bryan, K. M., Einarson, B., Kendall, R. L., Kolacz, K. H. & Saito, K. (1987). Specificity of cotranslational amino-terminal processing of proteins in yeast. *Biochemistry* **26**, 8242-8246.

Innis, M. A., Holland, M. J., McCabe, P. C., Cole, G. E., Wittman, V. P., Tal, R., Watt, K. W. K., Gelfand, D. H., Holland, J. P. & Meade, J. H. (1985). Expression, glycosylation, and secretion of an *Aspergillus* glucoamylase by *Saccharomyces cerevisiae*. *Science* **228**, 21-27.

Izumoto, Y., Sako, T., Yamamoto, T., Yoshida, N., Kikuchi, N., Ogawa, M. & Matsubara, K. (1987). Expression of human pancreatic secretory trypsin inhibitor in *Saccharomyces cerevisiae*. *Gene* **59**, 151-159.

Jayaram, M., Li, Y. -Y. & Broach, J. R. (1983). The yeast plasmid 2μ circle encodes components required for its high copy propagation. *Cell* **34**, 95-104.

Jigami, Y., Muraki, M., Nobuhiro, H. & Tanaka, H. (1986). Expression of synthetic human-lysozyme gene in *Saccharomyces cerevisiae*: use of a synthetic chicken-lysozyme signal sequence for secretion and processing. *Gene* **43**, 273-279.

Jilg, W., Schmidt, M., Zoulek, G., Lorbeer, B., Wilske, B. & Deinhardt, F. (1984). Clinical evaluation of a recombinant hepatitis B vaccine. *The Lancet* **ii**, 1174-1175.

Johnston, M. & Davis, R. W. (1984). Sequences that regulate the divergent *GAL1-GAL10* promoter in *Saccharomyces cerevisiae*. *Molecular and Cellular Biology* **4**, 1440-1448.

Johnston, M. (1987). A model fungal gene regulatory mechanism: the GAL genes of *Saccharomyces cerevisiae*. *Microbiological Reviews* **51**, 458-476.

Johnston, S. A. & Hopper, J. E. (1982). Isolation of the yeast regulatory gene *GAL4* and analysis of its dosage effects on the galactose/melibiose regulon. *Proceedings of the National Academy of Sciences, USA* **79**, 6971-6975.

Jones, E. (1984). The synthesis and function of proteases in *Saccharomyces cerevisiae*: genetic approaches. *Annual Review of Genetics* **18**, 233-270.

Julius, D., Blair, L., Brake, A., Sprague, G. & Thorner, J. (1983). Yeast α factor is processed from a larger precursor polypeptide: the essential role of a membrane-bound dipeptidyl aminopeptidase. *Cell* **32**, 839-852.

Julius, D., Brake, A., Blair, L., Kunisawa, R. & Thorner, J. (1984). Isolation of the putative structural gene for the lysine-arginine-cleaving endopeptidase required for processing of yeast prepro-α-factor. *Cell* **37**, 1075-1089.

Kenny, E. & Hinchliffe, E. (1989). Patent Application No. EP 317254.

Knowlton, R. G. (1986). Copy number and stability of yeast plasmids. In *Maximising Gene Expression*, (ed. W. Reznikoff & L. Gold), pp. 171-194. Butterworths: Boston.

Kramer, R. A., DeChiara, T. M., Schaber, M. D. & Hilliker, S. (1984). Regulated expression of a human interferon gene in yeast: control by phosphate concentration or temperature. *Proceedings of the National Academy of Sciences, USA* **81**, 367-370.

Kurjan, J. & Herskowitz, I. (1982). Structure of a yeast pheromone gene (*MF*α): a putative α-factor precursor contains four tandem copies of mature α-factor. Cell **30**, 933-943.

Lemire, J. M., Willcocks, T., Halvorson, H. O. & Bostian, K. A. (1985). Regulation of repressible acid phosphatase gene transcription in *Saccharomyces cerevisiae*. *Molecular and Cellular Biology* **5**, 2131-2141.

Lopes, T. S., Klootwijk, J., Veenstra, A. E., van der Aar, P. C., van Heerikhuizen, H., Raue, H. A. & Planta, R. J. (1989). High copy number integration into the ribosomal DNA of *Saccharomyces cerevisiae*: a new vector for high-level expression. *Gene* **79**, 199-206.

Lue, N. F., Chasman, D. I., Buchman, A. R. & Kornberg, R. D. (1987). Interaction of *GAL4* and *GAL80* gene regulatory proteins *in vitro*. *Molecular and Cellular Biology* **7**, 3446-3451.

Ma, J. & Ptashne, M. (1987). The carboxy-terminal 30 amino acids of GAL4 are recognised by GAL80. *Cell* **50**, 137-142.

McAleer, W. J., Buynak, E. B., Maigetter R. Z., Wampler, D. E., Miller, W. J. & Hilleman, M. R. (1984). Human hepatitis B vaccine from recombinant yeast. *Nature* **307**, 178-181.

Mellor, J., Dobson, M. J., Roberts, N. A., Tuite, M. F., Emtage, J. S., White, S., Lowe, P. A., Patel, T., Kingsman, A. J. & Kingsman, S. M. (1983). Efficient synthesis of enzymatically active calf chymosin in *Saccharomyces*. *Gene* **24**, 1-14.

Mellor, J., Malim., M., Gull, K,, Tuite, M. F., McCready, S., Dibbayawan, T., Kingsman, S. M. & Kingsman, A. J. (1985). Reverse transcriptase activity and Ty RNA are associated with virus-like particles in yeast. *Nature* **318**, 583-586.

Mellor, J., Kingsman, A. J. & Kingsman, S. M. (1986). Ty, an endogenous retrovirus of yeast? *Yeast* **2**, 145-152.

Melnick, L. M. Turner, B. G., Puma, P., Price-Tillotson, B., Salvato, K. A., Dumais, D. R., Moir, D. T., Broeze, R. J. & Avgerinos, G. C. (1990). Characterisation of a nonglycosylated single chain urinary plasminogen activator secreted from yeast. *Journal of Biological Chemistry* **265**, 801-807.

Moehle, C. M., Aynardi, M. W., Kolodny, M. R., Park, F. J. & Jones, E. W. (1987). Protease B of *Saccharomyces cerevisiae*: isolation and regulation of the *PRB1* structural gene. *Genetics* **115**, 255-263.

Moir, D. T. & Dumais, D. R. (1987). Glycosylation and secretion of human alpha-1-antitrypsin by yeast. *Gene* **56**, 209-217.

Moir, D. T., Mao, J. -I., Duncan, M. J., Smith, R. A. & Kohno, T. (1985). Production of calf chymosin by the yeast *S. cerevisiae*. *Development in Industrial Microbiology* **26**, 75-85.

Murray, A. W. & Szostak, J. W. (1983). Pedigree analysis of plasmid segregation in yeast. *Cell* **34**, 961-970.

Nakao, J., Miyanohara, A., Toh-e, A. & Matsubara, K. (1986). *Saccharomyces cerevisiae PHO5* promoter region: location and function of the upstream activation site. *Molecular and Cellular Biology* **6**, 2613-2623.

Orr-Weaver, T. C. & Szostak, J. W. (1983). Multiple tandem plasmid integration in *Saccharomyces cerevisiae*. *Molecular and Cellular Biology* **3**, 747-749.

Osborne, B. I. & Guarente, L. (1989). Mutational analysis of a yeast transcriptional terminator. *Proceedings of the National Academy of Sciences, USA* **86**, 4097-4101.

Penttilä, M. E., André, L., Lehtovaara, P., Bailey, M., Teeri, T. T. & Knowles, J. K. C. (1988). Efficient secretion of two fungal cellobiohydrolases by *Saccharomyces cerevisiae*. *Gene* **63**, 103-112.

Petes, T. D. (1979). Yeast ribosomal DNA genes are located on chromosome XII. *Proceedings of the National Academy of Sciences, USA* **76**, 410-414.

Romanos, M. A., Makoff, A. J., Beesley, K. M., Rayment, F. & Clare, J. J. (1990). Expression of tetanus toxin fragment C in *Saccharomyces*: complete gene synthesis is required to eliminate multiple transcriptional terminations. *Yeast* **6**, S427.

Rothstein, S. J., Lahners, K. N., Lazarus, C. M., Baulcombe, D. C. & Gatenby, A. A. (1987). Synthesis and secretion of wheat α-amylase in *Saccharomyces cerevisiae*. *Gene* **55**, 353-356.

Rudolph, H. K., Antebi, A., Fink, G. R., Buckley, C. M., Dorman, T. E., LeVitre, J, Davidow, L. S., Mao, J. -I. & Moir, D. T. (1989). The yeast secretory pathway is perturbed by mutations in *PMR1*, a member of a Ca^{2+} ATPase family. *Cell* **58**, 133-145.

Sabin, E. A., Lee-Ng, C. T., Shuster, J. R. & Barr, P. J. (1989). High-level expression and *in vivo* processing of chimeric ubiquitin fusion proteins in *Saccharomyces cerevisiae*. *Bio/Technology* **7**, 705-709.

Sakai, A., Shimizu, Y. & Hishinuma, F. (1988). Isolation and characterisation of mutants which show an oversecretion phenotype in *Saccharomyces cerevisiae*. *Genetics* **119**, 499-506.

Santangelo, G. M., Tornow, J., McLaughlin, C. S. & Moldave, K. (1988). Properties of promoters cloned randomly from the *Saccharomyces cerevisiae* genome. *Molecular and Cellular Biology* **8**, 4217-4224.

Schekman R. & Novick, P. (1982). The secretory process and yeast cell-surface assembly. In *The Molecular Biology of the Yeast Saccharomyces. Metabolism and Gene Expression*, (ed. J. H. Strathern, E. W. Jones & J. R. Broach), vol. 42, pp. 361-393. Cold Spring Harbor Laboratory Press: Cold Spring Harbor, New York.

Schultz, L. D., Hofmann, K. J., Mylin, L. M., Montgomery, D. L., Ellis, R. W. & Hopper, J. E. (1987). Regulated overproduction of the *GAL4* gene greatly increases expression from galactose-inducible promoters on multi-copy expression vectors in yeast. *Gene* **61**, 123-133.

Scolnick, E. M., McLean, A. A., West, D. J., McAleer, W. J., Miller, W. J. & Buynak, E. B. (1984). Clinical evaluation in healthy adults of a hepatitis B vaccine made by recombinant DNA. *Journal of the American Medical Association* **251**, 2813-2815.

Sleep, D., Belfield, G. P. & Goodey, A. R. (1990). The secretion of human serum albumin from the yeast *Saccharomyces cerevisiae* using five different leader sequences. *Bio/Technology* **8**, 42-46.

Sleep, D., Belfield, G. P., Ballance, D. J., Steven, J., Jones, S., Evans, L. R., Moir, P. D. & Goodey, A. R. (1991b). The selection of *Saccharomyces cerevisiae* strains that over-express heterologous proteins. *Bio/Technology* **9**, 183-187.

Sleep, D., Ogden, J. E., Roberts, N. A. & Goodey, A. R. (1991a). Cloning and characterisation of the *Saccharomyces cerevisiae* glycerol-3-phosphate dehydrogenase (*GUT2*) promoter. *Gene*, in press.

Smith, M., Leung, D. W., Gillam, S. & Astell, C. R. (1979). Sequence of the gene for iso-1-cytochrome c in *Saccharomyces cerevisiae*. *Cell* **16**, 753-761.

Smith, R. A., Duncan, M. J. & Moir, D. T. (1985). Heterologous protein secretion from yeast. *Science* **229**, 1219-1224.

St. John, T. P. & Davis, R. W. (1981). The organisation and transcription of the galactose gene cluster of *Saccharomyces*. *Journal of Molecular Biology* **152**, 285-315.

Stanway, C., Mellor, J., Ogden, J. E., Kingsman, A. J. & Kingsman, S. M. (1987). The UAS of the yeast *PGK* gene contains functionally distinct domains. *Nucleic Acids Research* **15**, 6855-6873.

Struhl, K. (1986). Yeast promoters. In *Maximising Gene Expression*, (ed. W. Rezinkoff & L. Gold), pp. 35-78. Butterworths: Boston.

Struhl, K., Stinchcomb, D. T., Scherer, S. & Davis, R. W. (1979). High frequency transformation of yeast: autonomous replication of hybrid DNA molecules. *Proceedings of the National Academy of Sciences, USA* **76**, 1035-1039.

Suzuki, K., Ichikawa, K. & Jigami, Y. (1989). Yeast mutants with enhanced ability to secrete human lysozyme: isolation and identification of a protease-deficient mutant. *Molecular and General Genetics* **219**, 58-64.

Tanner, W. & Lehle, L. (1987). Protein glycosylation in yeast. *Biochimica et Biophysica Acta* **906**, 81-99.

Tsai, P. -K., Frevert, J. & Ballou, C. E. (1984). Carbohydrate structure of *Saccharomyces cerevisiae* mnn9 mannoprotein. *Journal of Biological Chemistry* **259**, 3805-3811.

Valenzuela, P., Medina, A. & Rutter, W. J. (1982). Synthesis and assembly of hepatitis B virus surface antigen particles in yeast. *Nature* **298**, 347-350.

Wood, C. R., Boss, M. A., Kenton, J. H., Calvert, J. E., Roberts, N. A. & Emtage, J. S. (1985). The synthesis and *in vivo* assembly of functional antibodies in yeast. *Nature* **314**, 446-449.

Zaret, K. S. & Sherman, F. (1982). DNA sequence required for efficient transcription termination in yeast. *Cell* **28**, 563-573.

Chapter 5

The molecular biology of *Trichoderma reesei* and its application to biotechnology

Merja Penttilä, Tuula T. Teeri, Helena Nevalainen & Jonathan K. C. Knowles

Trichoderma reesei is a filamentous mesophilic soft rot fungus belonging to the Fungi Imperfecti. It was isolated in South East Asia during the Second World War from badly degraded cotton army tents. The degradation was later discovered to be caused by the cellulose hydrolysing enzymes secreted by the fungus. Although *Trichoderma* produces a number of different extracellular hydrolases such as xylanases and amylolytic enzymes, cellulases have drawn most of the research attention. The cellulolytic enzyme system of *Trichoderma* is the best studied among the cellulolytic organisms and consequently serves as a model system for enzymatic cellulose degradation in general.

Several different cellulases are secreted by the fungus which act in synergy to hydrolyse efficiently crystalline cellulose to glucose (reviewed by Montenecourt, 1983; Enari & Niku-Paavola, 1987). It is not yet exactly known how cellulose hydrolysis is brought about, but it is now clear that although all the enzymes hydrolyse the β-1,4-linkage between the glucose units in cellulose molecules, they differ in their specific activities and substrate specificities. This probably reflects the physically complex nature of the solid substrate which undergoes changes in solubility and cellulose chain length during hydrolysis.

The enzymes have been classified as cellobiohydrolases (CBH) (also called exoglucanases) which by definition release cellobiose units from the end of the cellulose molecules, and endoglucanases (EG) which hydrolyse bonds in random fashion in the middle of the cellulose chain. The small oligosaccharides and cellobiose generated by the action of the endo- and exoenzymes are then hydrolysed to glucose by β-glucosidase. A clear difference between the exo- and endoglucanases is that the endoglucanases are capable of hydrolysing substituted cellulose derivatives while the exoglucanases do not possess this activity. However, in spite of extensive biochemical characterization the activities of individual cellulases or even their exact number is not yet completely resolved.

Table 5.1. Industrial uses of *Trichoderma* cellulases

starch processing
animal feed
grain alcohol fermentation
malting and brewing
extraction of fruit and vegetable juices
pulp and paper industry
textile industry

It is generally believed that *Trichoderma* produces two different cellobiohydrolases. The genes and cDNAs encoding CBHI and CBHII have been isolated and characterized (Shoemaker *et al.*, 1983; Teeri, Salovuori & Knowles, 1983; Chen, Gritzali & Stafford, 1987; Teeri *et al.*, 1987a & b). The total number of *Trichoderma* endoglucanases is still under discussion. The genes for the two major endoglucanases, EGI and EGII (previously called EGIII), have been analyzed (Penttilä *et al.*, 1986; Van Arsdell *et al.*, 1987; Teeri *et al.*, 1987a; Saloheimo *et al.*, 1988).

Cellulase production is induced in cellulose-containing medium and repressed when *Trichoderma* is provided with an easily metabolizable carbon source such as glucose or glycerol (reviewed by Beguin, 1990; Bisaria & Mishra,1989). Under inducing conditions the cellulases make up over 90% of all the extracellular proteins secreted by the fungus. Mutagenesis and screening procedures have resulted in isolation of hypercellulolytic strains (reviewed by Nevalainen *et al.*, 1990b) some of which secrete over 40 g of cellulases into one litre of culture medium (Durand, Clanet & Tiraby, 1988; Durand *et al.*, 1988). About 60% of the secreted protein consists of the major cellobiohydrolase CBHI (Gritzali & Brown, 1979; Niku-Paavola, 1983).

Trichoderma strains are well adapted to fermenter cultivations and can utilize cheap raw materials for growth. The hyperproducing strains are used to produce cellulases in industrial scale for applications in which hydrolysis of plant material is needed. Industrial applications of *Trichoderma* cellulases are listed in Table 5.1.

The application of the techniques of molecular biology and genetic engineering to *Trichoderma* has opened up completely new possibilities to obtain knowledge of the organism and to improve its utilization in biotechnical applications. The existing data concerning gene structure and regulation have enabled modification of *Trichoderma* strains for production of novel mixtures of hydrolytic enzymes and even proteins of heterologous origin. Furthermore, the combination of genetic and protein engineering is leading to a better understanding of the function of the cellulolytic enzymes than was possible before.

Structure and function of *Trichoderma* cellulases

Sequence comparison of the cloned cellulase genes

Comparison of the primary structures deduced from the DNA sequences of the four *Trichoderma* cellulase genes cloned, *cbh1*, *cbh2*, *egl1* and *egl2* (previously called *egl3*), and later comparisons with other cellulases revealed an interesting common architecture. Cellulose degrading enzymes are composed of two structurally and functionally distinct domains; a large generally non-conserved core protein of approximately 500 amino acid residues and a small, so called tail domain, linked to either the N- or to the C- terminus of the core by a 30-40 amino acid residues long hinge peptide rich in serine, threonine and proline (Knowles, Lehtovaara & Teeri, 1987) (Fig. 5.1). Among *T. reesei* cellulases, the tail domains of approximately 30 amino acid residues are well conserved so that their amino acid sequence identity is over 70%. In the case of the core proteins only CBHI and EGI share significant sequence similarity (45%); CBHII and EGII are both clearly different in terms of amino acid sequence.

Comparison of the deduced protein sequences of these and other cellulases reveals that CBHII is similar to three bacterial endoglucanases of *Cellulomonas fimi*, *Microbispora bispora* and *Streptomyces* sp. (West *et al.*, 1989; Rouvinen *et al.*, 1990; Henrissat & Mornon, 1990). EGII is similar in sequence to at least one endoglucanase of *Schizophyllum commune* (Saloheimo *et al.*, 1988). Using the *Trichoderma cbh1* gene as a probe, the corresponding genes from *Humicola grisea* (Azevedo & Radford, 1990), *Phanerochaete chrysosporium* (Sims, James & Broda, 1988) and *Neocallimastis frontalis* (P. Reymond, R. Durand & M. Fevre, unpublished) have been isolated which shows that these genes share considerable sequence similarity.

Cellulase function as studied by limited proteolysis

Detailed characterization of the properties of fungal cellulases have long been hampered by the lack of pure enzymes. This is due to the fact that a large number of enzymes very similar in structure and other physical characteristics are induced simultaneously in the same culture conditions and their separation by conventional enzyme purification procedures has proved difficult. Fortunately, new approaches for the purification and analysis of cellulases have been developed recently (Van Tilbeurgh *et al.*, 1988; Tomme *et al.*, 1988a). In particular, efficient purification of the cellobiohydrolases is now possible using highly specific affinity chromatography (Tomme *et al.*, 1988a).

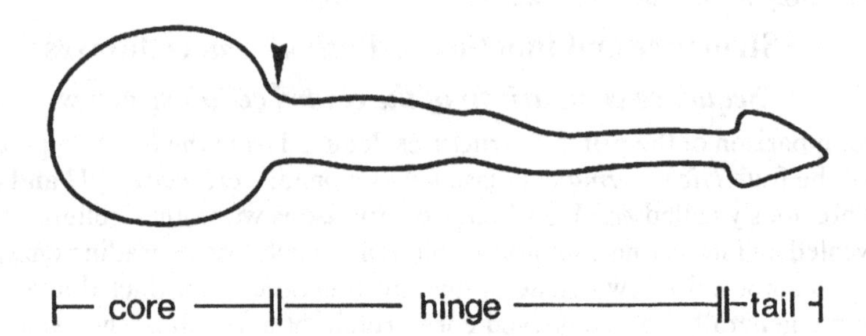

Fig. 5.1. Schematic structure of *Trichoderma* cellulases. Proportions correspond approximately to the data obtained by small angle X-ray diffraction for cellobiohydrolases (Schmuck *et al.*, 1986; Abuja *et al.*, 1988). The arrow marks the position of the proteolytic cleavage site (see text).

Limited proteolysis of *T. reesei* CBHI and CBHII has been shown to remove the hinge-tail region leaving the compact globular core protein essentially in tact (Tomme *et al.*, 1988b). Characterization of the properties of the core proteins revealed that their activities on small soluble oligosaccharides are comparable to those of the corresponding native enzymes (Van Tilbeurgh *et al.*, 1986; Tomme *et al.*, 1988b). Their activities and adsorption on insoluble polymeric cellulose were, however, dramatically reduced upon the removal of the tail. It was therefore suggested that the core domain contains the active site and the tail domain is required for interaction of the enzymes with crystalline cellulose.

The shapes of the intact and core proteins of CBHI and CBHII have been studied by small angle X-ray diffraction (Schmuck *et al.*, 1986; Abuja *et al.*, 1988). Analysis of the intact cellulases revealed a large globular protein with a long apparently flexible extension at one end. After limited proteolysis the terminal extension disappeared indicating that it corresponds to the tail-hinge region. The dimensions of the terminal region suggest that it has an almost fully extended conformation (Fig. 5.1).

Structure determination of cellulase domains

Genetic and protein engineering offer powerful tools for structure-function studies of enzymes but their efficient application requires the three dimensional structures of the enzymes to be determined. Crystallization of intact fungal cellulases has not so far been successful but the structure of the CBHII core protein has recently been solved (Rouvinen *et al.*, 1990) (Fig. 5.2). In addition, the NMR structure of a synthetic tail peptide of CBHI is also available (Kraulis *et al.*, 1989).

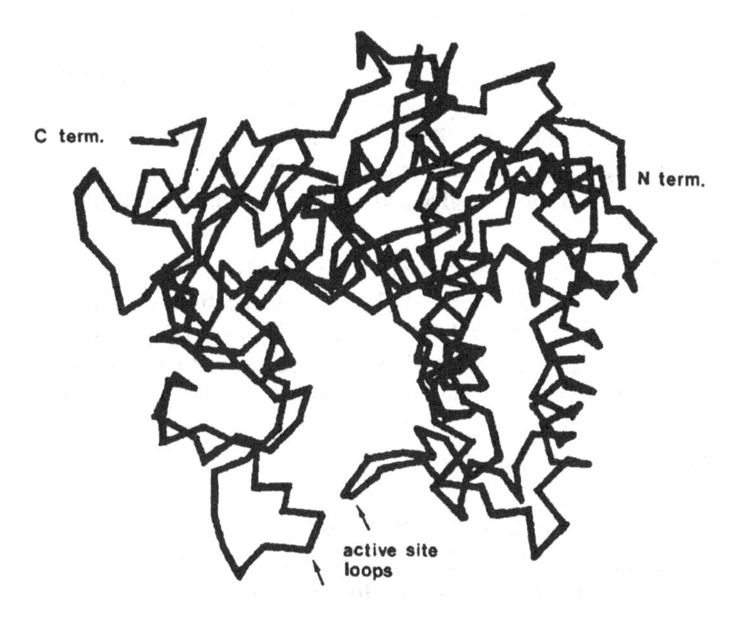

Fig. 5.2. Crystal structure of the CBHII core domain.

The CBHII core protein folds to a regular α/β barrel structure similar to that of triose phosphate isomerase (TIM) (Rouvinen *et al.*, 1990). This is a common fold in protein structures and is also found in other carbohydrase structures such as those of lysozymes and amylases (Matsuura *et al.*, 1984; Buisson *et al.*, 1987). However, instead of eight β- strands, which is typical of TIM, the CBHII structure consists of seven strands. In addition, CBHII only contains five α-helices while TIM has six.

Another remarkable feature of the CBHII core protein structure is the architecture of the active site. Analysis of the structure with different inhibitors and ligands diffused into the crystal revealed that the active site is located in a long tunnel formed by two stable surface loops (Rouvinen *et al.*, 1990) (Fig. 5.2). The four subsites for substrate binding, first proposed by Van Tilbeurgh *et al.* (1985), can be seen clearly in the tunnel. The unusual active site actually explains many of the biochemical properties of CBHII. For example, it is now evident that CBHII cannot hydrolyse soluble substituted cellulose simply because there is no space for the bulky substituents in the active site.

Comparison of the amino acid sequences of CBHII and the three bacterial endoglucanases also suggests a possible structural explanation for the differences in the enzymatic activities between endoglucanases and exoglucanases (Rouvinen *et al.*, 1990). In CBHII sequence there are several peptides which are clearly missing from the endoglucanase sequences. The most pronounced of these deletions in the endoglucanase sequences are located exactly in the position of the loops which close the active site tunnel in CBHII. Therefore, it seems possible that the endoglucanases might possess more open active site clefts or grooves perhaps facilitating degradation of glycosidic bonds in the middle of cellulose chains.

The tail domains are located at either the N- or the C-termini of the core domains of the different cellulases. The structure of CBHII reveals that both termini are located on the same side of the protein, opposite to the active site. Provided that the hinge region has sufficient flexibility, the tail domains may be positioned in similar orientations towards the active site in spite of their different relative positions in the amino acid sequence (Rouvinen *et al.*, 1990).

The structure of the tail domain reveals a wedge shaped molecule with one face hydrophilic and the other more hydrophobic (Kraulis *et al.*, 1989). A striking feature of the hydrophilic side of the wedge is the presence of three tyrosine residues forming a regular flat surface. Tyrosines have been shown to be important in substrate binding sites of many carbohydrate recognizing enzymes (Quiocho, 1986) and the modification of tyrosines has also been shown to reduce the binding of CBHI on crystalline cellulose (Claeyssens & Tomme, 1990). It is therefore likely that these tyrosines are central for the binding events. These and other hypotheses based on the structural data are currently being tested by site directed mutagenesis in our laboratory.

The yeast *Saccharomyces* is a convenient host for production of mutated *Trichoderma* cellulases for structure-function analysis. The cellulases are efficiently secreted into the yeast culture medium and are enzymatically active (Penttilä *et al.*, 1987a, 1988; Penttilä, Lehtovaara & Knowles, 1989) and the production levels are sufficient for large scale purification of the enzymes (Zurbriggen *et al.*, 1990a & b). In addition, using the techniques of genetic engineering, novel *Trichoderma* strains have been constructed in which a specific cellulase gene has been inactivated (see next section). These strains are well suited for the production of mutant forms of that particular enzyme. The ability to express the altered proteins in the native host is especially useful when the authenticity of the product, e.g. in glycosylation, is considered important.

Modification of cellulase production

T. reesei produces a complete set of cellulases required for efficient hydrolysis of crystalline cellulose to glucose. This is advantageous in traditional applications in which complete hydrolysis of the substrate for the production of, for example, single cell protein or ethanol is desired. More recently it has become important to modify lignocellulose specifically by selected or limited hydrolysis. For these new applications, strains producing novel cellulase profiles would be desirable.

Using mutagenesis and screening procedures it has so far been difficult to isolate *Trichoderma* strains deficient in only one or two particular cellulolytic enzymes, although hypercellulolytic (reviewed by Nevalainen *et al.*, 1990b) as well as cellulase negative (Nevalainen & Palva, 1978; Shoemaker, Raymond & Bryner, 1981; Durand *et al.*, 1988) strains have been isolated. However, genetic engineering can be used for the construction of strains devoid of, or with increased activity of, a single enzyme if the corresponding genes have been isolated.

A number of different transformation procedures have been developed for *T. reesei*. Using these methods DNA can be introduced into a range of previously unmodified strains as well as into strains already transformed earlier. The *Aspergillus amdS* gene (Penttilä *et al.*, 1987b) and the phleomycin resistance gene (Durand *et al.*, 1988; our unpublished results) can be used as dominant selection markers for the transformation of any *T. reesei* strain. Auxotrophic mutants of *Trichoderma* have been isolated which can be complemented with the corresponding *Aspergillus* genes *argB* (Penttilä *et al.*, 1987b) or *trpC* (E. Nyyssönen & H. Nevalainen, unpublished). Recently, complementation of a *Trichoderma* mutant with the *Neurospora* and the *Aspergillus pyr* gene has been reported (T. Berges, C. Barreau & J. Begueret, personal communication; Gruber *et al.*, 1990; M. Ward, personal communication). As in other filamentous fungi, cotransformation frequencies are high in *Trichoderma* (Penttilä *et al.*, 1987b) which can be important for easy strain construction.

Strains producing increased amounts of EGI

Analyses of the steady state mRNA levels indicate that the promoter of the *cbh1* gene is the strongest of the four cellulase genes analyzed (A. Jokinen & M. Penttilä, unpublished). To study whether the amount of EGI could be increased when produced under the control of the *cbh1* promoter, the *egl1* cDNA was linked to this promoter and the expression cassette was transformed into *T. reesei* by cotransformation with a plasmid carrying the *amdS* selection marker (Harkki *et al.*, 1990; Nevalainen *et al.*, 1990 a & b; A. Mäntylä &H. Nevalainen, unpublished), and as a linear fragment that carried the *amdS* gene (T. Karhunen & P. Suominen, unpublished) (Fig. 5.3).

Several of the endoglucanase-producing transformants obtained from the cotransformation experiment were analyzed. Interestingly, in the best producer strain, the expression cassette had been integrated into the chromosomal *cbh1* locus. This strain produced two to three fold higher

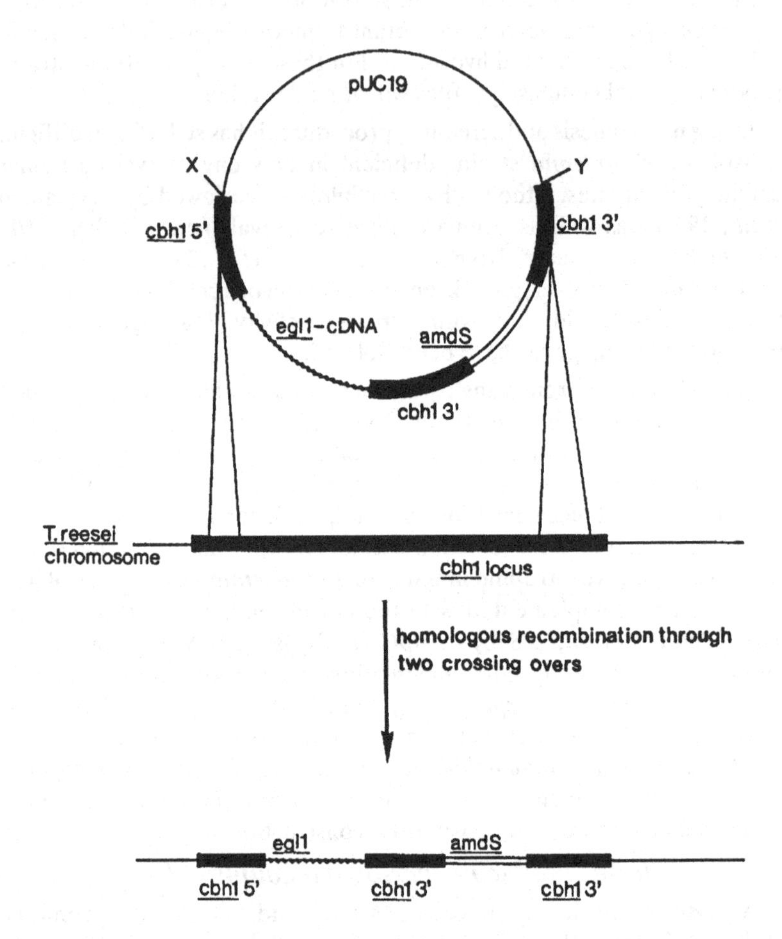

Fig. 5.3. Simultaneous gene replacement and introduction of the *egl1* gene into the *cbh1* locus of *Trichoderma*. The plasmid can be linearized before transformation by cutting at X and Y.

amounts of EGI compared to the original strain (Harkki *et al.*, 1990). The same plasmid was also cotransformed into a *T. reesei* strain in which the *cbh1* gene had been previously inactivated. Also in this case the best EGI producer had the expression cassette integrated into the *cbh1* locus. Southern analysis revealed that integration had occurred through the *cbh1* terminator sequences present in the transforming DNA (Harkki *et al.*, 1990). One copy of the integrated *egl1* gene resulted in a greater than four-fold increase in the proportion of EGI in the extracellular protein mixture, as measured using monoclonal antibodies. A similar improvement in EGI production was obtained when the *egl1* cDNA was introduced into *T. reesei* on a linear molecule which replaced the chromosomal *cbh1* locus in integration (T. Karhunen & P. Suominen, unpublished) (Fig. 5. 3).

These *Trichoderma* enzyme preparations in which EGI is enriched to comprise over 30% of the total secreted protein are useful in biotechnical processes utilizing grain such as barley and oats. Endoglucanases are mainly responsible for the hydrolysis of β-glucans and these new enzyme preparations give better hydrolysis of grain processed for monogastric animal feed.

Inactivation of cellulase genes

Trichoderma strains lacking either of the two cellobiohydrolases CBHI or CBHII have been obtained by insertional inactivation (Harkki *et al.*, 1990; Nevalainen *et al.*, 1990a & b) or by direct gene replacement. For instance, simultaneous inactivation of chromosomal *cbh1* gene by targeted introduction of a cDNA copy of the *egl1* gene into this locus is an efficient way to increase the proportion of endoglucanases in the cellulase preparation as discussed earlier.

To construct *T. reesei* strains deficient in production of CBHII, plasmids carrying the *A. nidulans argB* or *trpC* gene in between the 5' and 3' flanking regions of the *cbh2* gene were constructed. These were transformed into a *T. reesei argB* or *trpC* mutant respectively, selecting for complementation of the auxotrophic mutation (P. Suominen, A. Mäntylä & H. Nevalainen, unpublished). In a reasonable number of the transformants the integrating DNA had replaced the endogenous *cbh2* gene.

Monoclonal antibodies are of crucial importance for rapid and straightforward screening of the inactivation of a specific cellulase gene. Several transformants can be studied for the lack of, for instance, CBHII in the culture supernatant using CBHII specific monoclonal antibodies and the inactivation of the gene can be verified subsequently by Southern analysis of only a few transformants.

The novel enzyme preparations lacking CBHI or CBHII are of special interest in the paper and pulp industry. Gene replacement can now be used to inactivate other genes in successive transformations of the already modified strains to provide further variation in the enzyme preparations produced by *Trichoderma*.

Trichoderma as a production host for heterologous proteins

Industrial strains of *Trichoderma* have the ability to secrete several tens of grams of cellulases per litre of culture medium. It is noteworthy that in the hyperproducing strains studied, there is only one copy of each of the four cellulase genes, *cbh1*, *cbh2*, *egl1* and *egl2*, in the genome (our unpublished results). This suggests that the promoters of the cellulase genes, especially that of *cbh1*, function very efficiently to provide the high enzyme yields. The expression level of this promoter is increased at least 2000 fold in cellulose containing medium compared to repressed conditions (e.g. media containing glucose). The promoter of the *Trichoderma* phosphoglycerate kinase gene (*pgk*) (Vanhanen *et al.*, 1989) involved in glycolysis, is clearly weaker than the *cbh1* promoter as judged by the steady state mRNA levels (A. Jokinen, M. -L. Onnela & M. Penttilä, unpublished).

Although very little is yet known about the post-translational modifications of proteins in filamentous fungi it is anticipated that, as eukaryotes, they would be capable of carrying out the modifications necessary for the production of, for example, animal proteins with authentic properties. There are indications that *Trichoderma* will not hyperglycosylate proteins in a similar manner to the yeast *Saccharomyces*. For example, CBHI produced by *Trichoderma* contains N-glycans consisting of five to nine mannose residues (Salovuori *et al.*, 1987) while the same enzyme produced in *Saccharomyces* is heavily overglycosylated as are the other three cellulases, CBHII, EGI and EGII (Penttilä *et al.*, 1987a, 1988; Van Arsdell *et al.*, 1987).

The good production properties in large scale fermenter cultivations as well as the availability of strong promoters and techniques of genetic engineering make *Trichoderma* a promising candidate for production of heterologous proteins. Foreign proteins have been expressed in *Trichoderma* using both the *cbh1* and the *pgk* promoters. As expected on the basis of the RNA levels of the endogenous genes, higher yields of heterologous proteins have so far been obtained with the *cbh1* promoter (our unpublished results).

The DNA sequences coding for heterologous proteins have been linked to the *Trichoderma* expression vectors in between the promoter and terminator regions of the *cbh1* gene. As cotransformation frequencies are high (Penttilä *et al.*, 1987b) the expression cassette can be transformed into the fungus together with a separate plasmid carrying the selection marker for transformation. Transformation with a linear fragment increases targeted integration and replacement of the *cbh1* locus although non-homologous integration also occurs frequently.

Expression of calf chymosin

Chymosin is an aspartyl proteinase used in large amounts in cheese manufacture. The isolation of this product from the stomachs of young calves is expensive and alternative ways of production have been studied extensively.

Chymosin is naturally synthesized as preprochymosin. The signal sequence is cleaved off upon secretion resulting in prochymosin which is

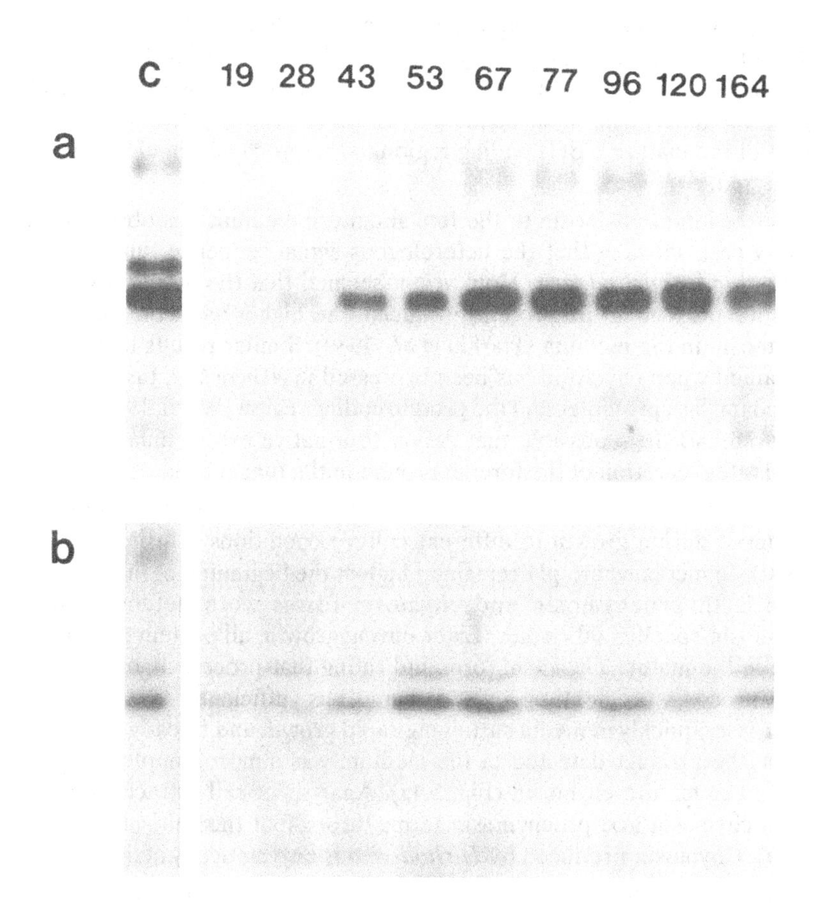

Fig. 5.4. Chymosin produced by *Trichoderma* (strain VTT-D-88361, Harkki *et al.*, 1989) in whey extract medium in a bioreactor (Uusitalo *et al.*, 1990). The cultivation time (h) is marked above the lanes. Panel a shows chymosin detected by antibodies in culture supernatants and panel b chymosin in cell extracts. Purified chymosin is shown as a control (c); upper band is prochymosin, lower band mature chymosin.

---

autocatalytically cleaved to the active chymosin form in acidic conditions (pH 4·5 or lower). If the pH is even lower (pH 2), prochymosin is cleaved to another active form, pseudochymosin which is 15 amino acids longer than chymosin. The fact that mature chymosin is only obtained after post-translational processing of the molecule allows different fusion proteins to be expressed and their production and secretion studied. For this purpose, several fusions of prochymosin cDNA to the *Trichoderma cbh1* promoter and signal sequence were constructed (Harkki *et al.*, 1989). The cDNA encoding preprochymosin was directly coupled to the *cbh1* promoter or the prochymosin cDNA was linked to the signal sequence of the *cbh1* gene. Also, signal sequence fusions or fusions of the N-terminal part of the mature CBHI coding region to the prochymosin cDNA were analyzed.

Secretion of chymosin to the fungal culture medium was observed in every case showing that the heterologous signal sequence functions in secretion in *Trichoderma*. However, it seemed that the more CBHI sequence present in the protein produced, the higher were the yields of chymosin in the medium (Harkki *et al.*, 1989). Similar results have been obtained when chymosin has been expressed in *Aspergillus*, fused to the glucoamylase promoter and the protein coding region (Ward, 1989; Ward *et al.*, 1990). It is possible that fusion to a native extracellular protein facilitates secretion of the foreign protein in the fungal host.

Production and processing of chymosin in *Trichoderma* was followed in detail during growth in different culture conditions (Uusitalo *et al.*, 1990). In media where pH remained high at the beginning of the cultivation both prochymosin and chymosin forms were detected with chymosin-specific antibodies. Later during growth, all protein produced was in the mature chymosin form, indicating that processing of prochymosin occurs efficiently provided that pH is sufficiently low. The pH decreases quickly in media sustaining rapid growth and already after two days, the product detected in the medium was almost completely processed to mature chymosin (Fig. 5.4a). Analysis of cell extracts showed both chymosin and prochymosin forms throughout the cultivation (Fig. 5.4b). Chymosin produced by *Trichoderma* is enzymatically active.

It is obvious, however, that the yields of chymosin are much lower than the amounts of CBHI secreted by the fungus. Although most of the chymosin produced is secreted, especially in fermenter cultivations (Harkki *et al.*, 1989; Uusitalo *et al.*, 1990), the secretion efficiency is not comparable to that of the cellulases. Furthermore, the steady state mRNA levels of chymosin are significantly lower than those of the *cbh1* gene (Harkki *et al.*, 1989).

Improved production levels are often obtained by screening the expression levels in different strains, by increasing the copy number of the heterologous gene in the host, and very often also by conventional uv or chemical mutagenesis of the initial transformant strain. Using a protease negative strain as a host the amounts of calf chymosin produced by *Aspergillus* transformants have been subsequently raised to commercially viable levels by mutagenesis and screening (Ward, 1989). As the initial chymosin yields produced by *Trichoderma* (about 40 mg l^{-1}) without carrying out any further strain development compare very favourably with those reported for *Aspergillus* (Cullen *et al.*, 1987; Ward, 1989), it is likely that considerable increases in chymosin levels could also be achieved in *Trichoderma*.

It is interesting that in the best chymosin producer among the *Trichoderma* transformants analyzed, the chymosin expression cassette had been integrated into the *cbh1* locus and inactivated the *cbh1* gene (Harkki *et al.*, 1989). As discussed earlier, this also occurred in transformants showing highest EGI expression under the *cbh1* promoter. In addition, this same phenomenon is observed in analyzing the production of some other heterologous proteins in *Trichoderma*, expressed under the *cbh1* promoter (our unpublished results). These results show that the site of integration in the chromosome can be very important for high level expression of a heterologous protein and indicate that position effect might play a role in optimal expression of the *cbh1* gene in *Trichoderma*.

Conclusions

Trichoderma reesei has long been, and still is, the best studied cellulolytic organism. The techniques of genetic engineering have greatly increased our understanding, not only of the cellulase enzymes but also of the molecular biology of the fungus itself. We now have all the tools available for thorough investigation of the properties of this organism and for the utilization of this knowledge in present and future biotechnical applications. Already the methods of transformation, targeted integration and gene replacement have enabled construction of novel *Trichoderma* strains producing altered mixtures of cellulolytic enzymes which are better suited for industrial applications than the cellulase preparations obtained with the conventional production strains. It is now also possible to obtain exact molecular data concerning the regulation of cellulase gene expression which can be later used to improve the biotechnical properties of the fungus.

The study of the cellulase genes of *Trichoderma* has provided us with the basis for interesting hypotheses concerning the function of the corresponding enzymes. However, for rational structure-function studies a three dimensional structure of the protein is needed. The structure determination of the catalytic domain of *Trichoderma* CBHII and the tail domain of CBHI, the first structures of cellulolytic enzymes ever determined, already give answers to some of the questions concerning the action of cellulase enzymes. More importantly, it is now possible to make precise hypotheses about the role of certain amino acids for the function of the enzyme and test these by site directed mutagenesis.

Filamentous fungi have recently acquired interest as production hosts for heterologous proteins. The availability of strong promoters and the good biochemical properties of *Trichoderma* make it a promising alternative host for protein production. Although the secretion of, for example, active calf chymosin and some other heterologous proteins, shows the potential of this fungus as a production host, much more information is needed to solve the problems, such as secretion inefficiency or mRNA instability which might reduce the yields obtained. In addition, almost nothing is known about post-translational processing of proteins in any filamentous fungus, which might furthermore differ significantly in different fungal species.

Cellulose is the most abundant renewable carbon source on earth; its efficient utilization and use of lignocellulosic material as a whole has acquired renewed interest. In addition to conventional biotechnical applications where complete hydrolysis of plant material is desired, processes which aim at partial hydrolysis or modification of lignocellulosic material are being developed. The use of enzymes offers new specificity and an environmentally friendly alternative to the harsh chemicals presently used. The combination of genetic and protein engineering will be crucial in providing essential information which allows us to utilize and modify the hydrolytic enzymes and fungal strains for future applications.

References

Abuja, P. M., Schmuck, M., Pilz, I., Tomme, P., Claeyssens, T. & Esterbauer, H. (1988). Structural and functional domains of cellobiohydrolase I from *Trichoderma reesei*. *European Biophysical Journal* **15**, 339-342.

Azevedo, M. de O. & Radford, A. (1990). Sequence of the *cbh-1* gene of *Humicola grisea* var. *thermoidea*. *Nucleic Acid Research* **18**, 668.

Beguin, P. (1990). Molecular biology of cellulose degradation. *Annual Review of Microbiology* **44**, 219-248.

Bisaria, V. & Mishra, S. (1989). Regulatory aspects of cellulase biosynthesis and secretion. *CRC Critical Reviews in Biotechnology* **9**, 61-103.

Buisson, G., Duèe, E., Haser, R. & Payan, F. (1987). Three dimensional structure of porcine pancreatic amylase at 2·9 Å resolution. Role of calcium in structure and activity. *EMBO Journal* **6**, 3909-3916.

Chen, C. M., Gritzali, M. & Stafford, D. W. (1987). Nucleotide sequence and deduced primary structure of cellobiohydrolase II of *Trichoderma reesei*. *Bio/Technology* **5**, 274-278.

Claeyssens, M. & Tomme, P. (1990). Structure-function relationships of cellulolytic proteins from *Trichoderma reesei*. In *Trichoderma Cellulases: Biochemistry, Genetics, Physiology and Applications*, (ed. C. P. Kubicek, D. E. Eveleigh, H. Esterbauer, W. Steiner & E. M. Kubicek-Pranz), pp. 156-167. The Royal Society of Chemistry: Cambridge.

Cullen, D., Gray, G. L., Wilson, L. J., Hayenga, K. J., Lamsa, M. H., Rey, M. W., Norton, S. & Berka, R. M. (1987). Controlled expression and secretion of bovine chymosin in *Aspergillus nidulans*. *Bio/Technology* **5**, 369-376.

Durand, H., Baron, M., Calmels, T. & Tiraby, G. (1988). Classical and molecular genetics applied to *Trichoderma reesei* for the selection of improved cellulolytic industrial strains. In *Biochemistry and Genetics of Cellulose Degradation*, ed. J. P. Aubert, P. Beguin, & J. Millet, pp. 135-151. Academic Press: San Diego.

Durand, H., Clanet, M. & Tiraby, F. (1988). Genetic improvement of *T. reesei* for large scale cellulase production. *Enzyme Microbial Technology* **10**, 341-345.

Enari, T. -M. & Niku-Paavola, M. -L. (1987). Enzymatic hydrolysis of cellulose: Is the current theory of the mechanism of the hydrolysis valid? *CRC Critical Reviews in Biotechnology* **5**, 67-87.

Gritzali, M. & Brown, R. D. Jr., (1979). The cellulase system of *Trichoderma*. The relationship between purified extracellular enzymes from induced or cellulose grown cells. *Advances in Chemistry Series* **181**, 237-260.

Gruber, F., Visser, J., Kubicek, C. P. & de Graaff, L. H. (1990). The development of a heterologous transformation system for the cellulolytic fungus *Trichoderma reesei* based on a *pyrG*-negative nmutant strain. *Current Genetics* **18**, 71-76.

Harkki, A., Mäntylä, A., Penttilä, M., Muttilainen, S., Bühler, R., Suominen, P., Knowles, J. K. C. & Nevalainen, H. (1990). Genetic engineering of *Trichoderma* to produce strains with novel cellulase profiles. *Enzyme Microbial Technology*, in press.

Harkki, A., Uusitalo, J., Bailey, M., Penttilä, M. & Knowles, J. K. C. (1989). A novel fungal expression system: secretion of active calf chymosin from the filamentous fungus *Trichoderma reesei*. *Bio/Technology* **7**, 596-603.

Henrissat, B. & Mornon, J. -P. (1990). Comparison of *Trichoderma* cellulases with other β-glycanases. In *Trichoderma Cellulases: Biochemistry, Genetics, Physiology and Applications*, (ed. C. P. Kubicek, D. E. Eveleigh, H. Esterbauer, W. Steiner & E. M. Kubicek-Pranz), 12-29. The Royal Society of Chemistry: Cambridge.

Knowles, J. K. C., Lehtovaara, P. & Teeri, T. T. (1987). Cellulase families and their genes. *Trends in Biotechnology* **5**, 255-261.

Kraulis, P., Clore, G. M., Nilges, M., Jones, T. A., Pettersson, G., Knowles, J. K. C. & Gronenborn, A. M. (1989). Determination of the three dimensional structure of the C-terminal domain of cellobiohydrolase I from *Trichoderma reesei*. *Biochemistry* **28**, 7241-7257.

Matuura, Y, Kusunoki, M., Harada, W. & Kakudo, M. (1984). Structure and possible catalytic residues of Taka-Amylase A. *Journal of Biochemistry* **95**, 697-702.

Montenecourt, B. S. (1983). *Trichoderma reesei* cellulases. *Trends in Biotechnology* **1**, 156-161.

Nevalainen, H. & Palva, E. T. (1978). Production of extracellular enzymes in mutants isolated from *Trichoderma viride* unable to hydrolyze cellulose. *Applied Environmental Microbiology* **35**, 11-16.

Nevalainen, H., Penttilä, M., Harkki, A., Teeri, T. T. & Knowles, J. K. C. (1990a). The molecular biology of *Trichoderma* and its application to the expression of both homologous and heterologous genes. In *Molecular Industrial Mycology*, (ed. S. A. Leong & R. Berka), 129-148. Marcel Dekker Inc.: New York.

Nevalainen, H., Harkki, A., Penttilä, M., Saloheimo, M., Teeri, T. T. & Knowles, J. K. C. (1990b). *Trichoderma reesei* as a production organism for enzymes for the paper and pulp industry. In *Biotechnology in Pulp and Paper Manufacture*, (ed.

T. Kent Kirk & H. -M. Chang), pp. 577-583. Butterworth-Heinemann Publishers: Boston.

Niku-Paavola, M. -L. (1983). Biochemistry of *Trichoderma reesei*, VTT-D-80133 cellulases. In *Proceedings of the Soviet Union-Finland Seminar on Bioconversion of Plant Raw Materials by Micro-organisms,* (ed. L. A. Golovela & O. N. Okunev, pp. 16-31.Academy of Sciences of the U.S.S.R.: Tashkent,

Penttilä, M., Lehtovaara, P., Nevalainen, H., Bhikhabhai, R. & Knowles, J. K. C. (1986). Homology between cellulase genes of *Trichoderma reesei*: complete nucleotide sequence of the endoglucanase I gene. *Gene* **45**, 253-163.

Penttilä, M., André, L., Saloheimo, M., Lehtovaara, P. & Knowles, J. K. C. (1987a). Expression of two *Trichoderma reesei* endoglucanases in the yeast *Saccharomyces cerevisiae*. *Yeast* **3**, 175-185.

Penttilä, M., Nevalainen, H., Rättö, M., Salminen, E. & Knowles, J. K. C. (1987b). A versatile transformation system for the cellulolytic filamentous fungus *Trichoderma reesei*. *Gene* **61**, 155-164.

Penttilä, M., André, L., Lehtovaara, P., Bailey, M., Teeri, T. T. & Knowles, J. K. C. (1988). Efficient secretion of two fungal cellobiohydrolases in *Saccharomyces cerevisiae*. *Gene* **63**, 103-112.

Penttilä, M., Lehtovaara, P. & Knowles, J. K. C. (1989). Cellulolytic yeast strains and their application. In *Yeast Genetic Engineering*, ed. P. J. Barr, A. J. Brake & P. Valenzuela, pp. 247-267. Butterworths: Boston.

Quiocho, F. A. (1986). Carbohydrate-binding proteins: tertiary structures and protein-sugar interactions. *Annual Review of Biochemistry* **55**, 287-315.

Rouvinen, J., Bergfors, T., Teeri, T. T., Knowles, J. K. C. & Jones, A. (1990). The three dimensional structure of cellobiohydrolase II from *Trichoderma reesei*. *Science* **279**, 380-386.

Saloheimo, M., Lehtovaara, P., Penttilä, M., Teeri, T. T., Ståhlberg, J., Johansson, G., Pettersson, G., Claeyssens, M. Tomme, P. & Knowles, J. K. C. (1988). A new endoglucanase from *Trichoderma reesei*: the characterization of both gene and enzyme. *Gene* **63**, 11-21.

Salovuori, I., Makarow, M., Rauvala, H., Knowles, J. K. C. & Kääriäinen, L. (1987). Low molecular weight high-mannose type glucans in a secreted protein of the filamentous fungus *Trichoderma reesei*. *Bio/Technology* **5**, 152-156.

Schmuck, M., Pilz, I., Hayn, M. & Esterbauer, H. (1986). Investigation of cellobiohydrolase from *Trichoderma reesei* by small angle X-ray scattering. *Biotechnological Letters* **8**, 397-402.

Shoemaker, S. P., Raymond, J. C. & Bruner, R. (1981). Cellulases: Diversity amongst improved *Trichoderma* strains. In *Trends in the Biology of Fermentations for Fuels and Chemicals*, ed. A. E. Hollaender, pp. 89-109. Plenum Press: New York.

Shoemaker, S., Schweikart, V., Ladner, M., Gelfand, D., Kwok, S., Myambo, K. & Innis, M. (1983). Molecular cloning of exo-cellobiohydrolase from *Trichoderma reesei* strain L27. *Bio/Technology* **1**, 691-696.

Sims, P., James, C. & Broda, P. M. A. (1988). The identification, molecular cloning and characterisation of a gene from *Phanerochaete chrysosporium* that shows strong homology to the exo-cellobiohydrolase I gene from *Trichoderma reesei*. *Gene* **74**, 411-422.

Teeri, T. T., Salovuori, I. & Knowles, J. K. C. (1983). The molecular cloning of the major cellulase gene from *Trichoderma reesei*. *Bio/Technology* **1**, 696-699.

Teeri, T. T., Kumar, V., Lehtovaara, P. & Knowles, J. K. C. (1987a). Construction of cDNA libraries by blunt end ligation: efficient cloning of long cDNAs from filamentous fungi. *Analytical Biochemistry* **164**, 1-67.

Teeri, T. T., Lehtovaara, P., Kauppinen, S., Salovuori, I. & Knowles, K. J. C. (1987b). Homologous domains in *Trichoderma reesei* cellulolytic enzymes: gene sequence and expression of cellobiohydrolase II. *Gene* **51**, 43-52.

Tomme, P., McCrae, S., Wood, T. M. & Claeyssens, H. (1988a). Chromatographic separation of cellulolytic anzymes. *Methods in Enzymology* **160**, 187-192.

Tomme, P., Van Tilbeurgh, H., Pettersson, G., Van Damme, J. Vandekerchove, J., Knowles, J. K. C., Teeri, T. T. & Claeyssens, M. (1988b). Studies of the cellulolytic system of *Trichoderma reesei* QM 9414. Analysis of domain function in two cellobiohydrolases by limited proteolysis. *European Journal of Biochemistry* **170**, 575-581.

Uusitalo, J., Nevalainen, H., Harkki, A., Knowles, J. K. C. & Penttilä, M. (1990). Enzyme production by recombinant *Trichoderma reesei* strains. *Journal of Biotechnology* **6**, in press.

Van Arsdell, J. N., Kwok, S., Schweickert, V. L. Ladner, M. B., Gelfand, D. H. & Innis, M. A. (1987). Cloning, characterization, and expression in *Saccharomyces cerevisiae* of endoglucanase I from *Trichoderma reesei*. *Bio/Technology* **4**, 60-64.

Van Tilbeurgh, H., Pettersson, G., Bhikhabhai, R., De Boeck, H. & Claeyssens, M. (1985). Studies of the cellulolytic system of *Trichoderma reesei* QM 9414. Reaction specificity and thermodynamics of interactions of small substrates and ligands with the 1,4-β-glucan cellobiohydrolase II. *European Journal of Biochemistry* **148**, 329-334.

Van Tilbeurgh, H., Tomme, P., Clayessens, M., Bhikhabhai, R. & Pettersson, G. (1986). Limited proteolysis of the cellobiohydrolase I from *Trichoderma reesei*. *FEBS Letters* **204**, 223-227.

Van Tilbeurgh, H., Loontiens, F. G., De Bruyne, C. K. & Claeyssens, M. (1988). Fluorogenic and chromogenic glycosides as substrates and ligands of carbohydrases. *Methods in Enzymology* **160**, 45-59.

Vanhanen, S., Penttilä, M., Lehtovaara, P. & Knowles, J. K. C. (1989). Isolation and characterization of the 3- phosphoglycerate kinase gene (*pgk*) from the filamentous fungus *Trichoderma reesei*. *Current Genetics* **15**, 181-186.

Ward, M. (1989). Heterologous gene expression in *Aspergillus*. In *Proceedings of the EMBO-Alko Workshop on Molecular Biology of Filamentous Fungi*, ed H. Nevalainen & M. Penttilä, pp. 119-128. Foundation for Biotechnical and Industrial Fermentation Research: Helsinki.

Ward, M., Wilson, L. J., Kodama, K. H., Rey, M, W. & Berka, R. (1990). Improved production of chymosin in *Aspergillus* by expression as a glucoamylase-chymosin fusion. *Bio/Technology* **8**, 435-440.

West, C., Elzanowski, A., Yeh, L. -S. & Barker, W. C. (1989). Homologues of catalytic domains of *Cellulomonas* glucanases found in fungal and *Bacillus* glucosidases. *FEMS Microbiological Letters* **59**, 167-172.

Zurbriggen, B., Bailey, M. J., Penttilä, M., Poutanen, K. & Linko, M. (1990a). Pilot scale production of a heterologous *Trichoderma reesei* cellulase by *Saccharomyces cerevisiae*. *Journal of Biotechnology* **13**, 267-278.

Zurbriggen, B., Penttilä, M., Viikari, L. & Bailey, M. J. (1990b). Pilot scale production of a *Trichoderma reesei* endo-β-glucanase by brewer's yeast. *Journal of Biotechnology*, in press.

Chapter 6

Expression of heterologous genes in filamentous fungi

R. Wayne Davies

Filamentous fungal species have been chosen for development as hosts for the commercial production of heterologous proteins for three main reasons: their enormous intrinsic capacity for secretion of proteins into the extracellular milieu as part of their saprotrophic lifestyle; their widespread use in fermentation-based industries; and the availability of good genetic systems and a reasonable level of molecular biology for some species. Heterologous proteins may be derived from other fungal species, or from entirely unrelated organisms. A compilation of the published examples of heterologous gene expression in filamentous fungi is given in Table 6.1. All fungal systems developed to date incorporate secretion signals into the transformation vectors.

The possibility of achieving high-level secretion of mammalian proteins was the major driving force. In addition, filamentous fungi are proving to be very competitive for intracellular protein production where large amounts are needed cheaply, and also provide companies with a strategy for circumventing patent blocks to the production of certain proteins in the more standard expression systems. Two cost-effective, commercial systems have been developed in the six-year period (1984-1990) since development began. These are the *Aspergillus nidulans* system based on the *alc*A (alcohol dehydrogenase I) promoter and its positive regulator *alc*R, which was developed at Allelix Inc. (Gwynne *et al.*, 1987; Davies, 1990) and the *Aspergillus niger* var. *awamori* system using the homologous glucoamylase promoter (or that from different *A. niger* strains) which was developed at Genencor (Cullen *et al.*, 1987; Ward, 1990).

The basic procedures for genetic transformation, and the phenomena observed are similar for all filamentous fungal species and have been discussed in detail in chapter 1. A number of vectors have been further developed as expression cassettes. Those currently in use are not particularly sophisticated, consisting typically of a strong constitutive or regulatable fungal promoter, an efficient translation start, a signal peptide coding sequence where secretion is the goal, cloning sites for the heterologous gene, and sequences for transcriptional termination and polyadenylation. In all cases, except the vector system of *Mucor* (Van

Table 6.1. Published examples of heterologous gene expression in filamentous fungi

Species	Gene expressed	Promoter used[a]	Reference[b]
A. nidulans	bovine prochymosin	*A. niger glaA*	1
A. nidulans	human α2 interferon	*alcA*, *A. niger glaA*	2
A. nidulans	*Cellulomonas fimi* endoglucanase	*alcA*, *A. niger glaA*	2
A. nidulans	human tissue plasminogen activator	*alcC*, *tpiA*, *A. niger adhA*	3
A. oryzae	bovine prochymosin	*A. oryzae* α-amylase	4
Mucor circinelloides	*M. miehei* aspartyl proteinase	*M. miehei* aspartyl proteinase	5
A. oryzae	*Mucor miehei* aspartyl proteinase	*A. oryzae* α-amylase	6
A. nidulans	human epidermal growth factor	*alcA*	7
Trichoderma reesei	bovine prochymosin	*T. reesei cbh1*	8
A. nidulans	*E. coli* endotoxin subunit B	*amdS*	9
A. nidulans	hGMCSF	*adhA*	10
A. nidulans	bovine prochymosin	*oliC*	11

[a]See text for explanation of symbols; promoters not designated with a species of origin are all from *A. nidulans*. References: 1, Cullen *et al.*, 1987; 2, Gwynne *et al.*, 1987; 3, Upshall *et al.*, 1987; 4, Boel *et al.*,, 1987; 5, Dickinson *et al.*, 1987; 6, Christensen *et al.*, 1988; 7, Gwynne *et al.*, 1989; 8, Harkki *et al.*, 1989; 9, Turnbull *et al.*, 1989; 10, Upshall *et al.*, 1990; 11, Ward, 1990.

Heeswijk, 1986), stable transformants are only obtained by integration. This is advantageous in practice, since there is no need to maintain selection pressure on the transformants. The majority of integration events are usually at heterologous sites, with the homologous targeting frequency varying between 10% and 40% (occasionally as high as 80%) depending on the marker gene used and the extent of the homologous region. As originally shown by Tilburn *et al.* (1983), multi-copy transformants are common, and usually carry a head to tail tandem array of vector molecules at a single site, although more complex patterns also occur. Transformants with 10 copies of the vector are common, and strains with up to 50 copies can be found easily by screening Southern transfers. Multiple-copy transformants are mitotically stable in most cases (although thorough testing is advisable); certainly, *A. nidulans* transformants are sufficiently stable for extended fermentation runs.

In this chapter I shall present the basic features of the established commercial systems and certain others, discuss what has been learned about the biological basis of protein production by filamentous fungi and its limitations, and indicate where future research and development should be focused.

Expression-secretion systems in filamentous fungi

Aspergillus nidulans

A. nidulans was the species originally chosen for the development of expression-secretion systems by the three biotechnology companies that made serious attempts to exploit filamentous fungal systems; Allelix Inc. of Toronto, Genencor Inc. of San Francisco and Zymogenetics Corporation of Seattle. *A. nidulans* is related to both *A. niger* and *A. oryzae* which are the major mould species used in established industrial processes. However, unlike these organisms, it has both a sexual and parasexual cycle, and one of the most scientifically developed genetic systems available, centring around one original isolate from Glasgow (Pontecorvo *et al.*, 1953). Moreover, transformation was already established for *A. nidulans*, and, to this day remains considerably more efficient than transformation of other *Aspergillus* species, which is a factor in making *A. nidulans* the organism of choice for development work.

The Genencor team concentrated on utilising the promoter and secretion signal peptide coding region of the *A. niger* glucoamylase gene (*gla*A) which had already been cloned (Boel *et al.*, 1984), using *A. nidulans* as the initial host (Cullen *et al.*, 1987). The protein used for development work was bovine prochymosin. The Allelix team also used the *gla*A promoter and signal (Gwynne *et al.*, 1987, 1989) but concentrated their effort on using the *A. nidulans alc*A promoter linked to synthetic signal sequences, since the regulation of this promoter was well understood and could be manipulated. Human interferon $\alpha 2$ and a bacterial endoglucanase were the first proteins expressed at Allelix. At Zymogenetics, three promoters were examined: the ADH3 or *alc*C promoter of *A. nidulans* (from an alcohol dehydrogenase gene subject to ethanol but not glucose regulation which is probably not expressed in primary metabolism); the triosephosphate isomerase (*tpi*A) promoter of *A. nidulans*, a medium level constitutive promoter; and the high-level constitutive promoter of an alcohol dehydrogenase gene (*adh*A) from *A. niger.* Experimental experience with the Genencor, Allelix and Zymogenetics systems constitute the major body of knowledge in the field. The Allelix and Genencor systems were taken to full commercial application (see below), while the Zymogenetics filamentous fungal development programme was terminated after obtaining valuable initial results. The heterologous proteins expressed at Zymogenetics were human tissue plasminogen activator (tPA) (Upshall *et al.*, 1987) and human granulocyte-macrophage colony stimulating factor (h-GMCSF) (Upshall *et al.*, 1990). A tPA cDNA, modified to carry a *Bam*HI site on the 5′ side of the start codon, was fused to all three promoter-signal sequence vectors using the *tpi*A transcriptional terminator in all cases. The highest yields of secreted active tPA were obtained with the *adh*A promoter (13 to 300 μg l^{-1}), moderate yields with the *tpi*A promoter (3 to 47 μg l^{-1}), and the lowest yields with the *alc*C promoter in the presence of ethanol (0 to 7 μg l^{-1}), after 28 h growth at 37°C. When the *A. niger gla*A signal sequence and propeptide was fused to the first processing site of tPA, yields fell by about half. A Genencor *gla*A pre-pro construction gave a similar effect on prochymosin production (Cullen *et al.*, 1987). Clearly, efficient processing of proteins during and after secretion is very dependent on the correct structural context for the processing site. Yields of tPA of 2·5 mg l^{-1} were subsequently obtained. In the case of h-GMCSF, only the *adh*A promoter was used, and yields of secreted active protein of between 500 μg and 1 mg l^{-1} were obtained, but the protein was incorrectly cleaved at the signal peptide C-terminus.

Other promoters have been tried in *A. nidulans* in commercial development programmes, notably the *A. nidulans oli*C promoter from the nuclear gene encoding subunit 9 of the mitochondrial ATP synthase complex (Ward, 1990) and the promoter of the *A. niger* gene coding for acid phosphatase (*pac*A; MacRae *et al.*, 1988; Davies, 1990). It was reasoned that *oli*C would be expressed at its highest level during maximal logarithmic growth, thus improving on the *gla*A promoter which is expressed late in log phase. Despite this advantage in timing of expression, the steady-state level of chymosin mRNA produced in transformants was much lower than that of native *oli*C message, in contrast to *gla*A mRNA, so the promise of the system was not fulfilled (Ward, 1990). Similarly, the potentially convenient use of phosphate or pH regulation in expression systems in fermenters employing the *A. niger* acid phosphatase promoter was found not to be practicable commercially as the expression levels obtained were too low (W. D. MacRae, unpublished; Davies, 1990).

The *alc*A and *gla*A promoters remain the strongest available, although many other promoters of *A. nidulans* and related species have now been cloned and characterised, and are useful in academic contexts. For example, the *amd*S (acetamidase) promoter has been used to express the *Escherichia coli* enterotoxin subunit B gene in *A. nidulans* (Turnbull *et al.*, 1989). One alternative promoter system now in commercial use at Allelix employs a mutant form of the ethanol-regulated aldehyde dehydrogenase (*ald*A) promoter (Pickett *et al.*, 1987) which acts as a very high level constitutive promoter throughout growth (M. Devchand & D. I. Gwynne, personal communication).

Aspergillus niger var. *awamori*

A. niger var. *awamori* has become the standard host for commercial heterologous protein production at Genencor, using the *A. niger* or *A niger* var. *awamori glaA* promoter. Allelix scientists devised the first transformation system for *A. niger* (Buxton, Gwynne & Davies, 1985). A simple positive and negative selection system using mutants in the ATP sulphurylase gene was developed, which allowed any expression construction to be transformed into any production strain (including polyploids) of *A. niger* or many other filamentous fungi (Buxton, Gwynne & Davies, 1989).

Aspergillus oryzae

The promoter of the α-amylase gene cloned from an *A. oryzae* mutant expressing high levels of α-amylase has been used, with an *A. niger glaA* transcriptional terminator, for expression of a cloned *Mucor miehei* aspartyl proteinase in *A. oryzae* (Christensen *et al.*, 1988). Yields of up to 3 g l^{-1} were obtained in fermenters. It should be noted that it is fairly easy to produce transformants yielding g l^{-1} levels of heterologous fungal proteins (including in *A. nidulans*). When *A. oryzae* was used for chymosin production (Boel *et al.*, 1987), levels identical to those obtained in other *Aspergillus* species (10 mg l^{-1}) were obtained.

Trichoderma reesei

T. reesei has been used as the host for bovine chymosin production, using the promoter and transcriptional terminator of the highly-expressed *cbh* 1 (exo-cellobiohydrolase I) gene (Harkki *et al.*, 1989), and introducing the construct using cotransformation with the *amdS* gene from *A. nidulans* into acetamide non-utilising strains (Penttilä *et al.*, 1987; chapter 5). The steady-state levels of chymosin were low, and as with other systems, the level of chymosin secretion was much lower than that of fungal enzymes. Nevertheless, the levels of 40 mg l^{-1} achieved compared favourably with typical initial levels of 2 to 10 mg l^{-1} in *A. nidulans* and *A. awamori*. However, the genetics of *Trichoderma* is very undeveloped, and experience at Genencor has shown certain transformants to be unstable (Ward, 1990). *T. reesei* would seem only to be competitive with *Aspergillus* systems for the production of fungal proteins, or where adequate levels of a foreign protein can be achieved in the first step, so that no research-based strain improvement programme is necessary.

Other filamentous fungi

The Zygomycete fungus *Mucor circinelloides* was used as the host for expression of the aspartyl proteinase of *Mucor miehei* (Dickinson *et al.*,

1987), using the pMA67 *Mucor-E. coli* shuttle vector. This vector was one of the few examples of apparent autonomous replication in filamentous fungi (Van Heeswijk, 1986). Expression of the *Mucor miehei* gene was driven by its own promoter, the vector providing replication functions and a selectable marker (*leu*). Despite the potential high copy number of this vector, the yield among twelve transformants was only 1 to 12 mg l^{-1} of culture. The requirement for a constant selection pressure makes this system, as it stands, unlikely to find wide commercial application.

Neurospora crassa was the first filamentous fungus to be transformed (Case *et al*., 1979) but has not been used for heterologous gene expression, except for recent unpublished reports of the secretion of bovine chymosin. *Achlya ambisexualis*, an oomycete, has been transformed using G418 (geneticin) resistance. The expression of human γ-interferon from an SV40 promoter and herpes simplex thymidine kinase from the mouse metallothionein promoter were reported (Leung, Jing & Leung, 1987; Leung, 1987). Finally, transformation systems are available for the two major antibiotic-producing fungi *Penicillium chrysogenum* (Cantoral *et al*., 1987; Beri & Turner, 1987) and *Cephalosporium acremonium* (Skatrud *et al*., 1987). The latter has been used to increase cephalosporin production by introducing an extra copy of the gene for the ring expanding enzyme (Queener *et al*., 1988), but no reports of heterologous protein expression in these species have been published.

The *Aspergillus* commercial production systems

Development strategies

Initially, both the Allelix and the Genencor filamentous fungal expression-secretion systems were developed in *A. nidulans*. After non-optimal components of the constructions used were dealt with (the removal of amino acids created at junctions, the generation of precise signal peptide-coding region junctions, and the optimisation of the sequence upstream of the translational start site; Gwynne *et al*., 1987), the Allelix *alc*A system produced 1 mg l^{-1} of human interferon $\alpha 2$ and up to 20 mg l^{-1} of a bacterial endoglucanase. Genencor made four different fusions of the *gla*A and bovine prochymosin genes; the best initial levels of secreted chymosin (3 mg l^{-1}) were obtained with a precise fusion of prochymosin to the *gla*A promoter and signal peptide coding sequence (Cullen *et al*., 1987). These secretion levels were at least ten-fold better than the initial levels obtained somewhat later by Zymogenetics for tPA (see above). With this more promising start, both Allelix Inc. and Genencor Inc. persevered in developing systems that would reach commercial levels of production. The strategies adopted were different, and were determined by the scientific options open. Some of the expression vectors that have been developed are shown in Fig. 6.1.

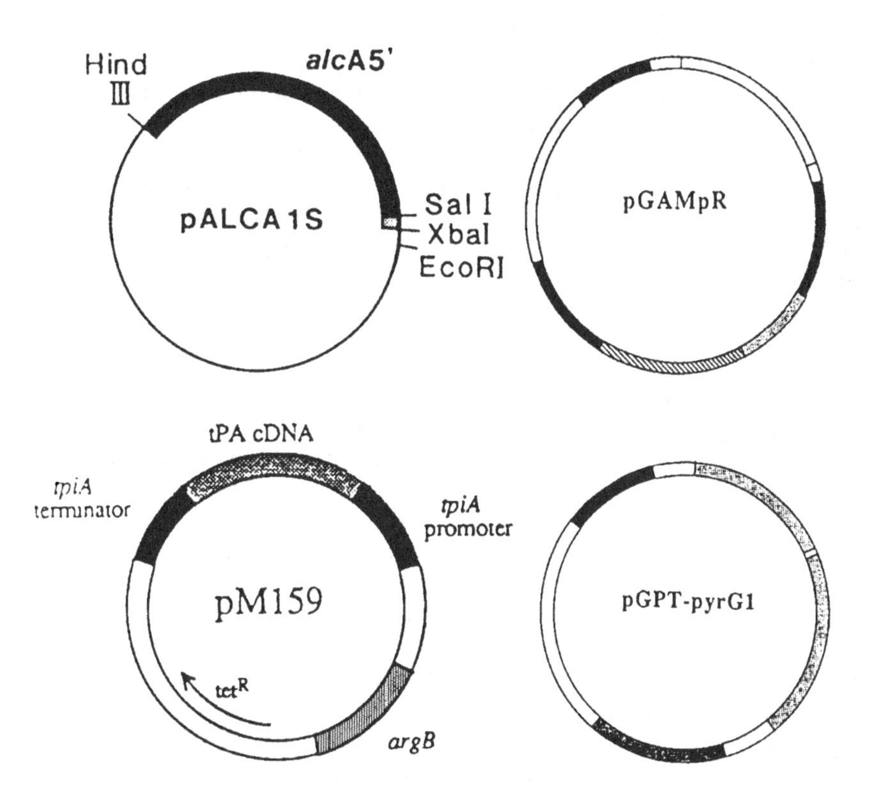

Fig. 6.1. Important vectors used in filamentous fungal expression system development. Clockwise from top left, they are: pALCA1S, one of the important early vectors used at Allelix Inc. (single line = bacterial vector sequence, shaded region = signal sequence; Gwynne *et al.*,, 1987a); pGAMpR, the glucoamylase complete gene-prochymosin fusion vector (shaded areas from top right clockwise indicate; *N. crassa pyr4*, *A. niger gla*A terminator, bovine prochymosin B cDNA, *A. awamori gla*A coding region, *A. awamori gla*A promoter, β-lactamase gene of pBR322; Ward, 1990); pGPT-pyrG1, a standardised Genencor expression vector (shaded areas from top right clockwise indicate; *A. niger gla*A terminator, *A. awamori gla*A promoter, *A. nidulans pyr*G gene, β-lactamase of pBR322); tPA expression vector pM159 of Zymogenetics (Upshall *et al.*, 1987).

The *alc*A promoter was part of a well-understood control system, so that it was possible potentially to manipulate both the promoter and its positive regulator. Thus, commercial development at Allelix remained in *A. nidulans* and concentrated on utilising the control system, with a strong emphasis on growth optimisation and fermentation to increase yields. Approaches were developed in parallel that allowed access to any strain of a related *Aspergillus* species, should a different host prove necessary (Buxton *et al.*, 1985, 1989). On the other hand, the gene(s) controlling *gla*A expression were unknown. Therefore, although this promoter is subject to the same control in *A. nidulans* as in *A. niger* (Gwynne *et al.*, 1987; Cullen *et al.*, 1987), without more basic research strain development at Genencor could not concentrate on raising transcription levels. Instead, development was switched to *A. awamori*, in the hope of encountering a background giving improved secretion performance, and focused on the interaction of proteins with the secretion system and in removing proteinases as a source of product degradation.

The alcA/alcR systems

The ethanol regulon of *A. nidulans* comprises two structural genes, *alc*A and *ald*A and one positive-acting regulatory gene, *alc*R (Sealy-Lewis & Lockington, 1984) which also stimulates its own synthesis. All three genes are directly subject to glucose repression mediated by the product of the *cre*A (carbon repression) gene (Bailey & Arst, 1975). When multicopy transformants of heterologous genes (or homologous genes) under *alc*A promoter control were produced, two interesting observations were made. Although there was wide variation in protein secretion levels at any given copy number there was an observable proportionality below about 5 to 10 integrated copies but not thereafter; in addition, the strains became allyl alcohol resistant. Allyl alcohol sensitivity is dependent upon expression of ADHI, which converts allyl alcohol into the toxic product acrolein. This result showed that the intrinsic *alc*A gene was not expressed at normal levels. The probable, though still formally unproven, basis for this effect was titration of *alc*R product by multiple *alc*A promoters. *A. nidulans* transformants such as T580 containing at least 50 copies of the *alc*R gene (under control of its own promoter) were therefore produced and used as the background for all future transformations aimed at high-level expression of foreign genes. On this background, the approximate copy-number dependence extended to at least 20 copies of *alc*A- $\alpha 2$ interferon (Gwynne *at al.*, 1988), and consistently higher levels of transcription per integrated copy were observed with a series of heterologous genes compared to a single *alc*R background.

Thus, the system in use at Allelix Inc. consists of the assembly of *alc*A promoter, synthetic or *gla*A signal, direct fusion to the heterologous coding region by oligonucleotide mutagenesis, and the *gla*A terminator, all integrated in multiple copies with a multicopy *alc*R background. Considerable effort was put into scaling up production by optimising growth, induction and fermentation conditions (Smart, 1990). A range of inducers, all volatile organic chemicals such as ethylmethylketone, were tested. The loss of inducer from fermenters was controlled by lowering temperature, aeration and stirring rates. It also proved important to control the amount of shearing of the mycelium. With the inducible system, biomass was built up, and inducer was added once glucose was sufficiently depleted to lift glucose repression, in the presence of a non-repressing carbon source. This was not necessary with the constitutive modified *ald*A promoter (see above), which yielded very high production levels. The inducible *alc*A system yields 2 to 3 g l^{-1} of fungal proteins such as *A. niger* glucoamylase, and 100 mg l^{-1} (human epidermal growth factor) to 300 mg l^{-1} of secreted active mammalian proteins. The system has also proved extremely useful for intracellular synthesis of heterologous mammalian proteins at levels up to 500 mg l^{-1}.

The A. awamori glaA system

Although the level of secreted chymosin was only up to 3 mg l^{-1} in initial transformants with the *gla*A *A. nidulans* system, intracellular levels of chymosin appeared to be as high as 50 mg l^{-1}. It seemed clear that secretion rather than transcription or translation was the major area needing development. Since we have no knowledge of filamentous fungal secretion pathways, except by analogy with yeast, the approach taken was to use different hosts that were hoped to provide more efficient secretion, and to initiate a classical mutagenesis and screening approach to generate strains that were able to secrete more of the prochymosin synthesised internally. A strain of *A. niger* var. *awamori* was chosen as the host because it overproduces glucoamylase, to a level of approximately 5 g l^{-1} in shake flasks. In fact, the level of chymosin secreted from the very best transformant (7 mg l^{-1}) was not much better than that obtained with *A. nidulans* as a host. This is an interesting result which, along with the experiments in *Trichoderma* reported above, indicates that the structural or enzymatic components of the secretion pathway are not dramatically different quantitatively between 'hypersecretory' strains of filamentous fungi and the standard *A. nidulans* laboratory strain. The Genencor team also switched to the homologous *A. niger* var. *awamori gla*A promoter, which performed very similarly to the *A. niger gla*A promoter.

Three steps were taken, however, which were successful in raising the level of secreted chymosin (Ward, 1990). First, a programme of multiple rounds of mutagenesis with *N*-methyl-*N*'-nitro-*N*-nitrosoguanidine followed by screening of the highest-yielding transformant was undertaken. Several mutant strains with elevated levels of chymosin production were isolated. Second, gene-targeting was used to produce strains with null mutations in the gene for aspergillopepsin A, an aspartyl proteinase suspected of degrading recombinant proteins, and this resulted in a modest two-fold increase in overall yield (Berka *et al.*, 1990). Finally, a new fusion of the prochymosin coding sequence in frame to the entire *A. niger* var. *awamori* glucoamylase gene yielded transformants that produced up to 150 mg l^{-1} of active secreted chymosin in an *A. niger* var. *awamori* aspergillopepsin A plus background. Transfer of this construct to an aspergillopepsin A mutant background, combined with mutagenesis and screening, produced strains with still higher yields (Ward, 1990).

Biological parameters of heterologous protein production

The combined experience of the research and development groups working in this field allow a number of interim conclusions to be drawn about the biological parameters controlling heterologous protein expression and secretion in filamentous fungi, thus indicating where further research might produce results of industrial importance.

It is clear that attaining high copy numbers of recombinant constructs is important, despite some published statements to the contrary. Differences in chromosome location, rearrangement of integrated sequences and differences between constructs generate an intrinsic high variability. If a threshold effect occurs, for instance the amount of regulatory protein being limiting, (as is the case for the *alc*R regulator of the *alc*A promoter) or by problematic interaction with the secretion pathway, (as in most of the prochymosin constructs) the effect of increasing copy number will be obscured. Therefore it is important to provide adequate levels of regulatory proteins, as in the Allelix system with *alc*R, and to be aware that for some heterologous proteins production should be limited to intracellular synthesis.

It is equally clear that the fungal secretion pathway handles fungal and bacterial proteins much better than it handles mammalian proteins (see Gwynne *et al.*, 1988; Turnbull *et al.*, 1989). Filamentous fungi, unlike yeast, use mammalian signal peptides efficiently, so the poor secretion observed is not related to the signal sequence structure. The work on tPA secretion from *A. nidulans* shows that this species possesses processing enzymes that correctly recognise mammalian signals and cleave propeptides with correct specificity and reasonable efficiency (Upshall *et al.*, 1990). Nevertheless, the proteolytic cleavage of the signal peptide from the mature protein remains a likely candidate for improvement. It is sensitive to structural alteration as shown by the effects of including the *gla*A propeptide sequence beyond the signal (Cullen *et al.*, 1987; Upshall *et al.*, 1987). Experiments at Zymogenetics with h-GMCSF with or without N-linked glycosylation sites showed that a requirement for glycosylation had no effect on the level of secretion. Therefore, the increased levels of hypergylcosylation (the addition of large numbers of mannose units to the N-acetyl-glucosamine/mannose core) found in high-expressing strains compared to low-expressing strains was a secondary effect, probably due to slow transit through the secretory pathway. The successful secretion of prochymosin in complete glucoamylase gene-prochymosin fusions suggests that there is nothing about the structure of mammalian proteins *per se* that prevents reasonably efficient secretion once access to the correct secretion pathway is ensured; it also suggests that important differences between fungal and mammalian proteins reside within the apoprotein sequence. Although Genencor may have solved the problem for prochymosin by using long fusions, this in itself presents scientific and economic problems for the production of the majority of proteins. Proteins which are not self-processing like prochymosin will require cleavage by externally provided enzymes acting at engineered signals. More basic research is needed to elucidate the differences between fungal and mammalian proteins and their interaction with components of the secretion pathway.

The possibility of proteinase degradation has always been a concern in these systems. However, proteinase expression and secretion is under very stringent nitrogen metabolite control, and is almost completely blocked by using a high concentration of ammonium ions in the culture medium. Controlling the pH of the culture medium to between 6·5 and 7·0 is advisable. Strains mutant in *are*A, the positive activator for proteinase expression, grew well and were used at Zymogenetics. The two-fold difference in chymosin yield between proteinase deficient and wild-type strains of *A. awamori* indicates that this proteinase can play a role, albeit a small one.

Many secreted mammalian proteins carry sites for N- or O-linked glycosylation. To be commercially useful the proteins produced in heterologous systems should be identical to the natural form, including the substituent carbohydrate structures. Fungal proteins are also glycosylated by N- and O-linked addition (Boel *et al*., 1984; Salovuori *et al*., 1987) and under certain conditions it has been claimed that the pattern of carbohydrate addition may be similar to that found in mammalian cells (Salovuori *et al*., 1987). This refers principally to the fact that hyperglycosylation is not observed with well-secreted proteins under normal growth conditions, in contrast to yeast (Yip *et al*., 1988). Hyperglycosylation can be a problem, however, when secretion or growth is not optimal (Upshall *et al*., 1990). Before identity with the mammalian product is claimed, the correct glycosylation structures must be verified by NMR spectroscopy. Rigorous testing of protein function and antigenicity is also essential.

Conclusions

Many heterologous fungal and bacterial proteins can be expressed intracellularly or secreted from a number of filamentous fungal species in very large amounts (hundreds of mg l^{-1} to several g l^{-1}). Mammalian proteins are generally expressed at much lower levels (at least ten-fold less). Provided that the correct transcriptional signals are present, the limitation appears to be in transit through the secretion pathway, possibly in choice of the correct route. In the case of prochymosin, greatly enhanced secretion (17·5 fold) was obtained by fusing the heterologous gene onto a complete fungal structural gene that is normally well secreted. In the *alc*A-*alc*R system exploitation of high copy numbers of heterologous gene constructs was attained by providing adequate levels of a regulatory protein and by optimisation of growth and fermentation conditions. Commercial levels of protein secretion and intracellular synthesis can be reached by a combination of these approaches.

The established systems use *Aspergillus nidulans* or *A. niger* var. *awamori* as the host species. No significant difference in performance between *A. niger* var. *awamori* and *A. nidulans* is apparent and the performance of *A. oryzae* and *Trichoderma reesei* with secretion of mammalian proteins is also not significantly superior. Since further development of filamentous fungal expression systems will become increasingly dependent on scientific manipulation, there seems to be no justification for developing a range of genetically inaccessible species as hosts, and every justification for continuing the development of *A. nidulans* systems.

The majority of therapeutically important proteins are used in low amounts so that present systems are quite adequate for their production. Filamentous fungal systems offer a particular advantage where large quantities are needed for topical delivery rather than injection. For injectables, the major outstanding question is that of identity with natural proteins, in particular the precise nature of glycosylation structures added to heterologous proteins in filamentous fungi.

Heterologous proteins secreted from filamentous fungi are correctly processed, and are active and available in commercially attractive amounts. If the use of filamentous fungal hosts is to be expanded beyond a limited range of products determined by market niche and patent avoidance strategies, there is a need for:

- investigation of the structural elements of fungal proteins that allow them to be secreted more efficiently than mammalian proteins;
- the development of a reliable fusion protein cleavage system;
- precise characterisation of glycosylation structures;
- investigation of mRNA stability and translation.

In vitro systems for studying *Aspergillus* translation and translocation are available (Devchand *et al*., 1989). The long-term future of filamentous fungal expression systems in an industrial context depends upon a concerted molecular analysis of these areas, rather than continued reliance on the available systems and classical approaches to strain improvement.

References

Bailey, C. & Arst, H. N. Jr. (1975). Carbon catabolite repression in *Aspergillus nidulans*. *European Journal of Biochemistry* **51**, 131-160.

Beri, R. K. & Turner, G. (1987). Transformation of *Penicillium chrysogenum* using the *Aspergillus nidulans amdS* gene as a dominant selectable marker. *Current Genetics* **11**, 639-641.

Berka, R. M., Ward, M., Wilson, L. J., Hayenga, K. J., Kodama, K. H., Carlomagno, L. P. & Thompson, S. A. (1990). Molecular cloning and deletion of the gene encoding aspergillopepsin A from *Aspergillus awamori*. *Gene* **86**, 153-162.

Boel, E., Hansen, M. T., Hjort, I., Hoegh, I & Fiil, N. P. (1984). Two different types of intervening sequences in the glucoamylase gene from *Aspergillus niger*. *EMBO Journal* **3**, 1581-1585.

Boel, E., Christensen, T. & Woldike, H.F. (1987). European Patent Application 238023.

Buxton, F. P., Gwynne, D. I. & Davies, R. W. (1985). Transformation of *Aspergillus niger* using the *argB* gene of *Aspergillus nidulans*. *Gene* **37**, 207-214.

Buxton, F. P., Gwynne, D. I. & Davies, R. W. (1989). Cloning of a new bidirectionally selectable marker for *Aspergillus* strains. *Gene* **84**, 329-334.

Cantoral, J. M., Diez, B., Barredo, J. L., Alvarez, E. & Martin, J. F. (1987). High frequency transformation of *Penicillium chrysogenum*. *Bio/Technology* **5**, 494-497.

Case, M. E., Schweizer, M., Kushner, S. R. & Giles, N. H. (1979). Efficient transformation of *Neurospora crassa* by utilizing hybrid plasmid DNA. *Proceedings of the National Academy of Sciences, USA* **76**, 5259-5263.

Christensen, T., Woldike, H., Boel, E., Mortensen, S. B., Hjortshoej, K., Thim, L. & Hansen, M. T. (1988). High level expression of recombinant genes in *Aspergillus oryzae*. *Bio/Technology* **6**, 1419-1422.

Cullen, D., Gray, G. L., Wilson, L. J., Hayenga, K. J., Lamsa, M. H., Rey, M. W., Norton, S. & Berka, R. M. (1987). Controlled expression and secretion of bovine chymosin in *Aspergillus nidulans*. *Bio/Technology* **5**, 369-376.

Davies, R. W. (1990). Molecular biology of a high-level recombinant protein production system in *Aspergillus*. In *Molecular Industrial Mycology: Systems and Applications in Filamentous Fungi*, (ed. S. A. Leong & R. Berka), pp. 45-81. Marcel Dekker: New York.

Devchand, M., Gwynne, D. I., Buxton, F. P. & Davies, R. W. (1989). An efficient cell-free translation system from *Aspergillus nidulans* and *in vitro* translocation of pre-pro-α-factor across *Aspergillus* microsomes. *Current Genetics* **14**, 561-566.

Dickinson, L., Harboe, M., Van Heeswijk, K., Sloman, P. & Jepson, L. H. (1987). Expression of active *Mucor miehei* aspartic protease in *Mucor circinelloides*. *Carlsberg Research Communications* **52**, 243-252.

Gwynne, D. I., Buxton, F. P., Williams, S. A., Garven, S. & Davies, R. W. (1987). Genetically engineered secretion of active human interferon and a bacterial endoglucanase from *Aspergillus nidulans*. *Bio/Technology* **5**, 713-719.

Gwynne, D. I., Buxton, F. P., Gleeson, M. A. & Davies, R. W. (1988). Genetically engineered secretion of foreign proteins from *Aspergillus* species. In *Protein Purification: Micro to Macro*, (ed. R. Burgess), pp. 355-365. Alan R. Liss Inc.: New York.

Gwynne, D. I., Buxton, F. A., Williams, S. A., Silb, M., Johnstone, J. A., Buch, J. K., Guu, Z. -M., Drake, D., Westphal, M., & Davies, R. W. (1989). Development of an expression system in *Aspergillus nidulans*. *Biochemical Society Transactions* **17**, 338-340.

Harkki, A., Uusitalo, J., Bailey, M., Penttilä, M. & Knowles, J. K. (1989). A novel fungal expression system: secretion of active calf chymosin from the filamentous fungus *Trichoderma reesei*. *Bio/Technology* **7**, 596-603.

Leung, W. -C. (1987). Expression and secretion of human interferon gamma in filamentous fungus *Achlya ambisexualis*. *Abstracts, 19th Lunteren lectures on Molecular Genetics* F241(b).

Leung, W. -C., Jing, G. Z. & Leung, M. F. K. (1987). Characterisation of herpes simplex virus thymidine kinase activity synthesised in recombinant filamentous fungus *Achlya ambisexualis*. *Abstracts, 19th Lunteren lectures on Molecular Genetics* F24(b).

MacRae, W. D., Buxton, F. P., Sibley, S., Garven, S., Gwynne, D. I., Davies, R. W. & Arst, H. N. Jr. (1988). A phosphate-repressible phosphatase gene from *Aspergillus niger*: its cloning, sequencing and transcriptional analysis. *Gene* **71**, 339-348.

Penttilä, M., Nevalainen, H., Ratto, M., Salininen, E. & Knowles, J. K. (1987). A versatile transformation system for the cellulolytic filamentous fungus *Trichoderma reesei*. *Gene* **61**, 155-164.

Pickett, M., Gwynne, D. I., Buxton, F. P., Elliott, R., Davies, R. W., Lockington, R. A., Scazzocchio, C. & Sealy-Lewis, H. M. (1987). Cloning and characterisation of the *ald*A gene of *Aspergillus nidulans*. *Gene* **51**, 217-226.

Pontecorvo, G., Roper, J. A., Hemmons, L. M., MacDonald, K. D. & Bufton, A. W. J. (1953). The genetics of *Aspergillus nidulans*. *Advances in Genetics* **5**, 141-238.

Queener, S. W., Skatrud, P. L., Ingolia, T. D., Yeh, W. K., Tietz, A. & McGilvray, D. (1988). Recombinant studies in *Cephalosporium acremonium* and *Penicillium chrysogenum*. *Abstracts, ASM Conference on the Molecular Biology of Industrial Microorganisms*, 40.

Salovuori, I., Makarow, M., Rauvala, H., Knowles, J. R. & Kaarainen, L. (1987). Low molecular weight high-mannose type glycans in a secreted protein of the filamentous fungus *Trichoderma reesei*. *Bio/Technology* **5**, 152-156.

Sealy-Lewis, H. M. & Lockington, R. A. (1984). Regulation of two alcohol dehydrogenases in *Aspergillus nidulans*. *Current Genetics* **8**, 253-259.

Smart, N. J. (1990). Scaling up production of recombinant DNA products using filamentous fungi as hosts. In *Molecular Industrial Mycology: Systems and Application in Filamentous Fungi*, (ed. S. A. Leong & R. Berka), pp. 251-279. Marcel Dekker: New York.

Skatrud, P. L., Queener, S. W., Carr, L. G. & Fisher, D. L. (1987). Efficient integrative transformation of *Cephalosporium acremonium*. *Current Genetics* **12**, 337-348.

Tilburn, J., Scazzocchio, C., Taylor, G. G., Zabicky-Zissman, J. M., Lockington, R. A. & Davies, R. W. (1983). Transformation by integration in *Aspergillus nidulans*. *Gene* **26**, 205-221.

Turnbull, I. F., Rand, K., Willetts, N. S. & Hynes, M. J. (1989). Expression of the *Escherichia coli* enterotoxin subunit B gene in *Aspergillus nidulans* directed by the *amdS* promoter. *Bio/Technology* **7**, 169-174.

Upshall, A., Kumar, A. A., Bailey, M. C., Parker, M. D., Favreau, M. A., Lewison, K. P. L., Joseph, M. L., Maraganore, J. M. & McKnight, G. L. (1987). Secretion of active human tissue plasminogen activator from the filamentous fungus *Aspergillus nidulans*. *Bio/Technology* **5**, 1301-1304.

Upshall, A., Kumar, A. A., Kaushansky, K. & McKnight, G. L. (1990). Molecular manipulation of and heterologous protein secretion from filamentous fungi. In *Molecular Industrial Mycology: Systems and Applications in Filamentous Fungi*, (ed. S. A. Leong & R. Berka), pp. 31-44. Marcel Dekker: New York.

Van Heeswijk, R. (1986). Autonomous replication of plasmids in *Mucor* transformants. *Carlsberg Research Communications* **51**, 433-443.

Ward, M. (1990). Chymosin production in *Aspergillus*. In *Molecular Industrial Mycology: Systems and Applications in Filamentous Fungi*, (ed. S. A. Leong & R. Berka), pp. 83-105. Marcel Dekker: New York.

Yip, C. L., Welch, S. K., Gilbert, T. & MacKay, V. L. (1988). Cloning and sequencing of the *S. cerevisiae* MNN9 gene required for hyperglycosylation of secreted proteins. *Yeast* **4** (special issue), s457.

Chapter 7

Methylotrophic yeasts as gene expression systems

R. A. Veale & P. E. Sudbery

The advantages of using yeast as a vehicle for expressing heterologous proteins are well known. Compared to the use of *Escherichia coli*, the proteins are likely to be produced in a more authentic fashion due to the capacity of yeasts as eukaryotes to carry out the various post-translational modifications which are often necessary. A particular advantage of yeasts is their ability to secrete proteins; this can prevent the accumulation of unwanted intracellular concentrations of the product, and is necessary for glycosylation and frequently for correct folding and processing.

So far, most work on heterologous gene expression has concentrated on the bakers' yeast *Saccharomyces cerevisiae*. This is because of the extensive genetic, biochemical and, latterly, molecular studies which have enabled the construction of sophisticated vector systems for the introduction and expression of foreign genes. An additional factor has been the experience of large scale fermentation accumulated from baking and brewing. Finally, because of its ancient involvement in food and brewing it has GRAS status (Generally Regarded As Safe).

Many proteins have now been expressed in *S. cerevisiae* (for reviews see Hirsch, Suarez Rendueles & Wolf, 1989; King, Walton & Yarranton, 1989). Generally, powerful promoters are used so as to maximise the level of expression. Typically, these are derived from glycolytic genes such as *PGK* (phosphoglycerate kinase). While these can provide high levels of expression, the actual level is often much lower than that of the homologous gene product of the promoter. To overcome this problem it may be necessary to use multiple copies of the construct on replicating plasmids. This in turn can cause problems of gene and plasmid stability. A further difficulty caused by the use of these promoters, is that they often lack regulatory properties allowing the expression of the heterologous gene to be controlled. If the protein is toxic then cell growth is adversely affected and selection imposed for the loss of the gene or plasmid. Selection systems have been developed based on naturally regulated genes such as that of the *GAL1,10* promoter (induced by galactose) and the *PHO5* promoter (induced by low levels of inorganic phosphate). Besides being naturally less powerful, these promoters require either difficult conditions for induction or the addition of an expensive inducer. Recently, hybrid constructs have been developed which combine the power and regulatory properties of glycolytic promoters and other promoters respectively. An example of this approach is the insertion of mating type operators within the *PGK* promoter (Walton & Yarranton 1989). Conditional regulation is then effected by the use of a $sir3^{ts}$ allele.

Methylotrophic yeasts such as *Hansenula polymorpha* and *Pichia pastoris* can be used as an alternative system to express heterologous proteins. These yeasts retain the advantages of *S. cerevisiae* while offering additional features including the use of a powerful but easily regulated alcohol oxidase (*AOX*) promoter. *AOX* is sequestered in peroxisomes where it forms a crystalline lattice. It is responsible for the first step in methanol utilisation oxidising it to formaldehyde and hydrogen peroxide. The hydrogen peroxide is subsequently acted upon by catalase, being broken down to water and molecular oxygen. The formaldehyde can either be assimilated into cellular carbon, via dihydroxyacetone synthetase and a cyclical pathway, or dissimilated to carbon dioxide and water via formate in a linear oxidation pathway (for a review see Veenhuis Van Dijken & Harder, 1983).

During growth on methanol the *AOX* promoter is induced and expressed to the extent that alcohol oxidase amounts to about 30% of total cell protein. In the presence of ethanol or high concentrations of glucose (0·5%) alcohol oxidase is undetectable. The mechanism of regulation is a combination of glucose or ethanol repression and methanol induction. In *H. polymorpha* glucose derepression in a chemostat leads to about 25% of maximum activity, the addition of methanol is necessary for full induction. Mixed substrate growth with glucose/methanol or glucose/formate leads to full induction. The *AOX* gene of *P. pastoris* has an absolute methanol requirement for expression.

The properties of the *AOX* promoter offer considerable advantages over those used in *S. cerevisiae*. Its strength allows high level expression to be achieved from a single integrated copy which ensures stability during large scale culture. Its tight regulation allows precise control over its expression by simple manipulations to the growth conditions. Other advantages are offered by methylotrophic yeasts. Methanol is a cheap carbon source which may offer an advantage in the production of low value/high volume products such as food processing enzymes. Large scale fermentations have been carried out and high cell densities may be achieved (130 g dry weight l^{-1} in *P. pastoris*, Wegner 1983). Moreover, *H. polymorpha* at least is thermotolerant with a growth maximum of 42°C, simplifying cooling problems in large fermenters.

Expression systems have been developed in both *P. pastoris* and *H. polymorpha*. In the former, Tumour Necrosis Factor (TNF)(Sreekrishna

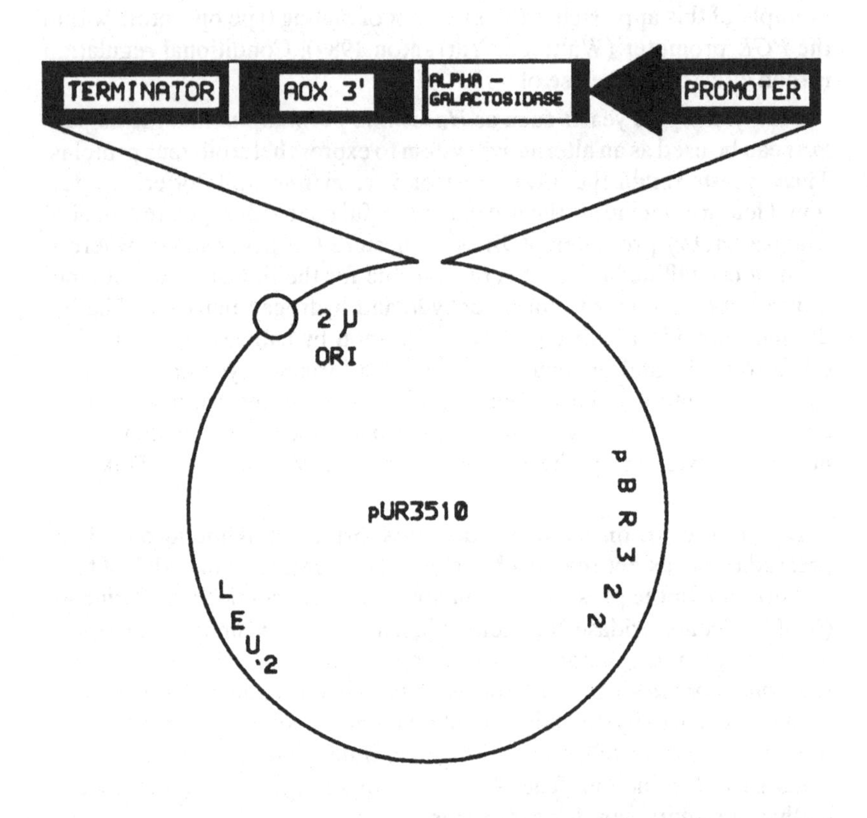

Fig. 7.1. Structure of pUR3510. A 5 kb expression cassette is shown inserted into the *Bam*H1 site of YEp13 (Broach, Strathern & Hicks, 1979) the components of the cassette are as follows: a) 1·6 kb 5' upstream sequences from the alcohol oxidase (*AOX*) gene (labelled 'promoter'); b) 131 bp synthetic oligonucleotide containing the *Saccharomyces cerevisiae* pre-invertase sequence fused in frame to the α-galactosidase coding sequence from the guar plant (*Cyamopsis tetragonobla*). The 5' terminus of the linker is joined to the *AOX* promoter sequences via a *Hgi*A1 site 33 bp upstream of the *AOX* ATG codon. The 5' sequences so omitted are replaced in the linker. At the 3' end the linker is fused to α-galactosidase ATG initiation codon via an *Nco*1 site; c) 1·1 kb coding region of the α-galactosidase sequence (labelled α-galactosidase); d) a 2·6 kb sequence from 3' terminus of the *AOX* region containing 1·3 kb of the 3' *AOX* coding sequence (labelled *AOX* 3') and 1·3 kb of downstream non-coding sequence (labelled terminator). The whole cassette was isolated as a *Bam*H1-*Hin*dIII fragment and cloned into the *Bam*H1 of YEp13 using synthetic linkers. YEp13 sequences are shown as a continuous line.

et al., 1989), hepatitis B surface antigen (HBsAg)(Cregg *et al.*, 1987), and invertase from *S. cerevisiae* (Tschopp *et al.*, 1987) have been expressed to high levels. In *H. polymorpha*, expression systems have been demonstrated by Janowitz *et al.* (1988), Shen *et al.* (1989) and by ourselves (Sudbery *et al.*, 1988). In this chapter we demonstrate the use of the system by the high level expression of α-galactosidase from the guar plant (*Cyamopsis tetragonobla*). This enzyme is used in food processing to convert guar gum (a food stabiliser) into a higher quality glue as a result of its function of cleaving the galactose side chains. Other α-galactosidases are unable to do this under the conditions used.

Expression of α-galactosidase

Construction of pUR3510

In order to express the α-galactosidase gene plasmid pUR3510 was constructed (Fig. 7.1). This is a replicating plasmid designed to express and secrete the α-galactosidase protein using the *AOX* promoter and terminator sequences. It consists of a 1·5 kb promoter fragment joined to the 1·1 kb α-galactosidase coding sequence via a 131 bp synthetic linker which contains the invertase secretion leader from *S. cerevisiae*. Translation is initiated from the invertase ATG codon and the α-galactosidase sequence fused in frame so that the mature protein is released after processing at the invertase cleavage sites. A 2·6 kb fragment containing the 3' coding sequence from the *AOX* gene and 1·3 kb of non-coding 3' sequences acts as a transcriptional terminator. The 5 kb expression cassette so formed was cloned into the *Bam*H1 site of YEp13. In *H. polymorpha* YEp13 replicates unstably as a multicopy plasmid (Gleeson, Ortori & Sudbery., 1986). This sequence also complements the *leu1-1* mutation of *H. polymorpha* allowing selection of transformants.

Construction of the A16 recipient strain

Initially, pUR3510 was transformed into the *H. polymorpha* strain *leu1-1*. Expression of α-galactosidase was detected, but levels were low (data not shown). Further investigation revealed that this was probably due to the host strain having a poor growth capacity on methanol compared to other strains of *H. polymorpha*, in particular strain CBS4732 which had been widely used in physiological experiments. As well as growing poorly, levels of the *AOX* product itself were reduced. It was decided to express the plasmid in a *H. polymorpha* strain that has the CBS4732 background. Such a strain would not, however, carry the *leu1-1* mutation necessary to select transformants; nor could the allele be introduced by genetic crossing because CBS4732 carries no genetic markers necessary to force the cross with the *leu1-1* strain. Any attempt to introduce a marker by mutagenesis ran the risk of altering the properties of the

CBS4732 strain by the introduction of other secondary mutations. Attempts were made to isolate spontaneous auxotrophic mutants; however, no mutants were recovered from 2×10^5 colonies which were screened.

In order to isolate an auxotrophic mutant without mutagenesis, suicide enrichment was used (Gleeson & Sudbery, 1988). The principle of this technique is to culture cells in a potentially selective condition such as minimal medium, in the presence of an agent which selectively kills growing cells, in this case we used nystatin. After treatment, survivors were plated out on non-selective medium and tested by replica plating to selective medium. Previous use of this technique has shown that it would be possible to recover spontaneous mutations which occurred at a rate that is too low to allow their isolation by screening alone. Using this technique on a culture of CBS4732, 21 auxotrophic colonies were recovered from 176 survivors of the nystatin treatment. If one mutant had been recovered from the screen for the spontaneous mutations then the enrichment factor of the suicide selection is $2 \cdot 4 \times 10^4$. All 21 mutants were found to have an adenine-requiring (Ade$^-$) phenotype upon characterisation. A bias towards this type of mutant occurring spontaneously has already been observed (Gleeson & Sudbery, 1988).

The mutants were tested for their ability to grow on methanol and considerable variation was found. One strain which showed good growth was selected and used to cross to the *leu1-1* strain. Leucine-requiring (Leu$^-$) progeny were selected and screened for their growth rate on methanol. A rapidly growing Leu$^-$ segregant from this cross was used to back-cross to the original Ade$^-$ CBS4732 strain. This process was repeated until a total of five crosses to the CBS4732 derivative had been completed. The final result was a strain, designated A16, which carried *leu1-1* but was over 95% isogenic to CBS4732 and grew well on methanol.

Transformation and integration of pUR3510 into strain A16

Plasmid pUR3510 was transformed into A16 to produce a strain designated A3. Transformants expressed α-galactosidase (see below) but the plasmid was extremely unstable in the absence of selection. A derivative was therefore sought where stabilisation results from chromosomal integration. The procedure used to effect integration was long term culture under selective conditions. Transformants were cultured in YEPD (a complex complete medium) at 18°C. Selection is exerted because *leu1-1* mutants are cold sensitive even on complete medium. After 63 generations, 3 clones were isolated which were found to be stably Leu$^+$. These were designated 8/1, 8/2 and 8/3 respectively. Southern hybridisation analysis showed that stabilisation had occurred by integration into the genome at the *AOX* locus, leaving an intact copy of the *AOX* locus enabling the transformants to grow on methanol. However, the pattern of

Table 7.1. Effect of culture conditions on the expression of α-galactosidase and alcohol oxidase

Carbon source	Dry cell weight mg l^{-1}	Yield of α-galactosidase		% Secreted	Alcohol oxidase activity[b]
		mg g^{-1} dry weight	% SCP[a]		
A16 Non-transformed control					
Glucose	5·2	0	0	-	1017
Glucose/methanol	5·6	0	0	-	2693
Glucose/formaldehyde	5·2	0	0	-	2194
8/1 Integrated					
Glucose	5·0	3·1	2·3	7	679
Glucose/methanol	5·6	12·5	7·0	60	2060
Glucose/formaldehyde	5·4	14·2	11·0	50	1776
Methanol	3·5	7·2	3·2	41	1719
8/2 Integrated					
Glucose	5·0	5·0	4·2	33	881
Glucose/methanol	6·3	12·6	7·0	47	2388
Glucose/formaldehyde	5·4	14·7	8·5	60	1762
Methanol	2·8	9·1	2·9	44	2143
Glucose/formaldehyde/ yeast extract	6·2	16·8	9·5	86	1500
A3 Replicating					
Glucose	4·6	7·2	3·4	50	587
Glucose/formaldehyde	5·0	19·7	9·2	70	1084
Glucose/formaldehyde/ yeast extract	6·3	7·7	3·3	82	1803

[a]SCP = soluble cell protein; [b]units of alcohol oxidase expressed as (μmole O_2 released min^{-1} ml^{-1}) g^{-1} dry weight.

---

hybridisation did not conform to a simple model and it is probable that multiple recombination events had occurred.

The stability of the Leu^+ phenotype of these strains was found to be absolute and no leu^- segregants were ever recovered. Indeed, integrants produced by a similar procedure in the *leu1-1* strain did not yield Leu^- derivatives even after the application of suicide selection to select for such colonies.

Expression of α-galactosidase

Strains were cultivated in a chemostat under carbon limitation with a variety of different carbon sources using a minimal medium base which would ensure selection in the case of the strain with the replicating construct. Addition of substrate (methanol, formaldehyde or glucose) caused an immediate decrease in dissolved oxygen tension, showing that cells were able to oxidise the substrate. Glucose/methanol and glucose/formaldehyde mixed substrate fermentations were also used since such regimes can lead to even greater induction of the *AOX* promoter than glucose limitation alone (Giuseppin, personal communication). The α-galactosidase levels were determined by measuring enzymic activity, by staining protein gels with Coomassie Blue and by Western blotting. Table 7.1 shows the yield of α-galactosidase obtained during growth under different culture conditions in terms of mg enzyme protein g^{-1} dry weight and as a percentage of soluble cell protein (SCP). Table 7.1 also shows the levels of alcohol oxidase activity.

During growth on glucose, alcohol oxidase levels were lower than those observed during mixed substrate growth. This is because glucose limitation results in derepression only, whereas the presence of methanol or formaldehyde causes further induction (Eggeling & Sahm, 1980; Egli, 1982). The α-galactosidase yields were also highest during mixed substrate growth. No α-galactosidase expression was detectable during batch growth in medium containing excess glucose, conditions which cause repression of *AOX* promoter (data not shown). The expression of α-galactosidase from the *AOX* promoter thus retains normal physiological control. The high level of expression in the replicating strain results from the selective conditions used to maintain the multicopy plasmid.

Under these conditions α-galactosidase was secreted but the proportion was variable, ranging from 6·5% (strain 8/1 growing on glucose) to 70% (A3 growing on glucose/formaldehyde). In general, higher values were found under inducing rather than derepressing conditions. The proportion of the enzyme secreted was found to be increased by addition of yeast extract to the medium. Thus, strain 8/2 growing on glucose/formaldehyde/yeast extract secreted 86% of the product and the yield was slightly increased to 9·5% of total soluble protein. The proportion secreted was also increased in the autonomously replicating strain A3; however, the total yield (i.e intracellular + secreted) was significantly decreased to 3% of total soluble protein. This was shown to be due to plasmid loss, as the addition of yeast extract relaxes selection for cells containing the plasmid.

The α-galactosidase protein was studied through Western blotting and Coomassie Blue staining. Using Western blotting, the molecular weight

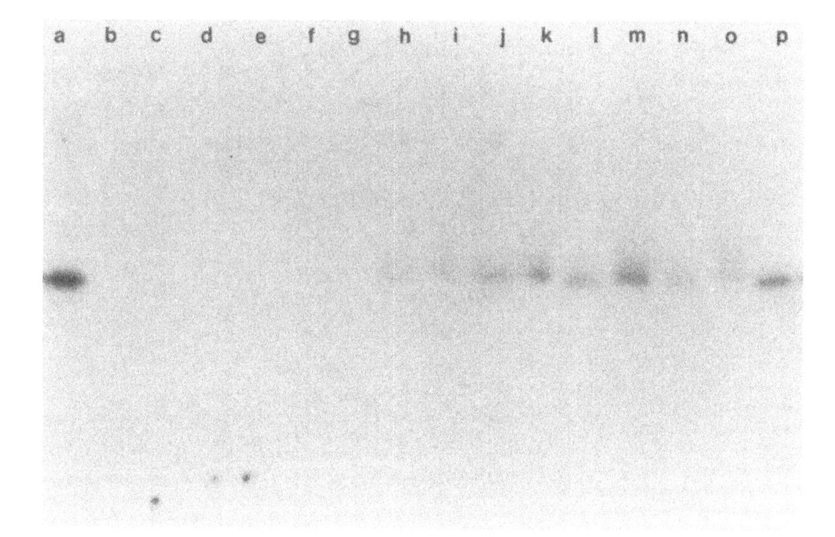

Fig. 7.2. Western blot of total cellular protein and samples from the culture supernatant from strains A16 (lanes b-g) and 8/2 (lanes h-o) grown on glucose (b, c, h, i), glucose/methanol (d, e, j, k), glucose/formaldehyde (f, g, l, m,) and methanol (n, o). Lysates (b, d, f, h, j, l, n) and supernatants (c, e, g, i, k, m, o) are arranged in pairs for each strain and growth condition. 25 ng of native α-galactosidase was loaded in lane a, and 8 ng of guar α-galactosidase produced in *S. cerevisiae* was loaded in lane p. Both are glycosylated.

of the product in both the intracellular and extracellular form could be compared to the native protein. The result is shown in Fig. 7.2. The product recovered from the lysate had a similar molecular weight to the native protein and to the protein produced by expression of the same gene in *S. cerevisiae*. Compared to the lysate, however, the material in the supernatant was of a slightly higher molecular weight and was heterogeneous in size, with material of higher molecular weight trailing back on the membrane. We interpret this as being due to glycosylation of the product, its diffuseness suggesting that the extent of glycosylation was variable.

Fig. 7.3 shows Coomassie Blue-stained proteins in the cell lysate and supernatant (concentrated 10-fold) from cells cultured under the different conditions described above. In the lanes containing material from the cell lysates the alcohol oxidase can be seen as a strong band of high molecular weight, its relative intensity reflecting the effects of the different growth conditions (derepression or induction). A band corresponding to α-galactosidase can be seen in the supernatant lanes, with the relative

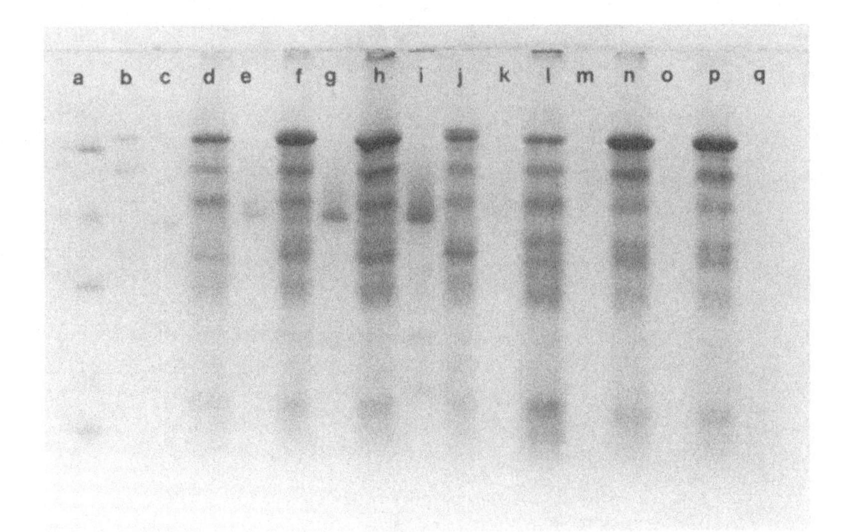

Fig. 7.3. Coomassie Blue stained polyacrylamide gels of total cell protein and material from the culture supernatant from strains 8/2 (lanes d-k) and A16 (lanes l-q). Cells were grown on glucose (d, e, l, m); glucose/methanol (f, g, n, o); glucose/formaldehyde (h, i, p, q) and methanol (j, k). Lysates (d, f, h, j, l, n, p,) and supernatants (e, g, i, k, m, o, q) are arranged in pairs for each growth condition and strain. The culture supernatant was concentrated 10-fold before loading. Lane a contains molecular weight standards (figures in kilodaltons); lane b contains 200 ng alcohol oxidase protein (the two bands represent monomeric and aggregated forms of the enzyme respectively); lane c contains 200 ng of native α-galactosidase.

intensities varying in the same way according to the growth conditions. This is in agreement with the estimates based on enzyme activities described above. The molecular weight of this band is slightly greater than the pure sample of guar α-galactosidase (lane c) probably due to glycosylation.

Conclusions

The high level expression of heterologous protein has now been demonstrated convincingly in both *H. polymorpha* and *P. pastoris*. In this chapter we have shown that it is possible to express guar plant α-galactosidase to a high level. The protein is both secreted and glycosylated. The protein is expressed from the *AOX* promoter and is under normal physiological control. When integrated into the chromosome it is completely stable.

References

Broach, J. R., Strathern, J. N. & Hicks, J. B. (1979). Transformation in yeast: development of a hybrid vector system and isolation of the *CAN1* gene. *Gene* **8**, 121-133.

Cregg, J. M., Tschopp, J. F., Stillman, C., Siegel, R., Akong, M., Craig, W. S., Buckholtz R. G., Madden, K. R., Kellaris, P. A., Davis, G. R., Smiley, B. L., Cruze, J., Terragrossa, R., Velicelebi, G. & Thill, G. (1987). High level expression and efficient assembly of hepatitis B surface antigen in the methylotrophic yeast *Pichia pastoris*. *Bio/Technology* **5**, 479-485.

Eggeling, L. & Sahm, H. (1980). Regulation of alcohol oxidase synthesis in *Hansenula polymorpha*: oversynthesis during growth on mixed substrates and induction by methanol. *Archives of Microbiology* **127**, 119-124.

Egli, T. (1982). Regulatory flexibility of methylotrophic yeasts in chemostat cultures: simultaneous assimilation of glucose and methanol at a fixed dilution rate. *Archives of Microbiology* **131**, 1-7.

Giuseppin, M. L. F., van EiJk, H. M. J. & Bes, B. C. M. (1988). Molecular regulation of methanol oxidase activity in continuous cultures of *Hansenula polymorpha*. *Biotechnology Bioengineering* **32**, 577-583.

Gleeson M. A. & Sudbery, P. E. (1988). Genetic analysis in the methylotrophic yeast *Hansenula polymorpha*. *Yeast* **4**, 293-303.

Gleeson, M. A., Ortori, G. S. & Sudbery, P. E. (1986). Transformation of the methylotrophic yeast *Hansenula polymorpha*. *Journal of General Microbiology* **132**, 3459-3465.

Hirsch, H. H., Suarez Rendueles, P., & Wolf, D. H. (1989). Yeast (*Saccharomyces cerevisiae*) proteinases: structure, characteristics and function. In *Molecular and Cell Biology of Yeasts*, ed. E. F. Walton & G. T. Yarranton, pp. 134-200. Blackie: Glasgow & London.

Janowicz, Z. A., Merckelbach, A., Eckart, M., Weydemann, U., Roggenkamp, R. & Hollenberg, C. P. (1988). Expression systems based on the methylotrophic yeast *Hansenula polymorpha*. *Yeast* **4**, S155.

King, D. J., Walton E. F. & Yarranton, G. T. (1989). The production of proteins and peptides from *Saccharomyces cerevisiae*. In *Molecular and Cell Biology of Yeasts*, ed. E. F. Walton & G. T. Yarranton, pp. 107-133. Blackie: Glasgow & London.

Shen, S. H., Bastien, L., Nguyen, T., Fung, M., Slilaty, S. N. (1989). Synthesis and secretion of hepatitis B surface antigen by the methylotrophic yeast *H. polymorpha*. *Gene* **84**, 303-309.

Sreekrishna, K., Nelles, L., Potencz, R., Cruze, J., Mazzaferro, P., Fish, W., Fuke, M., Holden, K., Phelps, D. & Wood, P. (1989). Expression, purification and characterisation of human tumour necrosis factor in the methylotrophic yeast *Pichia pastoris*. *Biochemistry* **28**, 4117-4125.

Sudbery, P. E., Gleeson, M. A., Veale, R. A., Ledeboer, A. M. & Zoetmulder, M. C. M. (1988). *Hansenula polymorphs* as a novel yeast system for the expression of heterologous genes. *Transactions of the Biochemical Society* **16**, 1081-1093.

Tschopp. J. F., Sverlow, G, Kosson, R., Craig, W. & Grinna, L. (1987). High level secretion of glycosylated invertase in the methylotrophic yeast *Pichia pastoris*. *Bio/Technology* **5**, 1305-1308

Veenhuis, M., Van Dijken, J. P. & Harder, W. (1983). The significance of peroxisiomes in the metabolism of one-carbon compounds in yeasts. *Advances in Microbial Physiology* **134**, 193-203.

Walton, E. F. & Yarranton, G. T. (1989). Negative regulation by mating type. In *Molecular and Cell Biology of Yeasts*, (ed. E. F. Walton & G. T. Yarranton), pp. 43-69. Blackie: Glasgow & London.

Wegner, E. H. (1983). Biochemical conversions by yeast fermentations at high cell densities. U.S. Patent number 4414329.

Chapter 8

Strain improvement of brewing yeast

Edward Hinchliffe

Brewing yeast have undergone no systematic or intentional breeding during their long history of use in the brewing industry. Traditionally, the brewers' choice of yeast strain has been based upon empirical observation of the yeast's fermentation performance and the flavour and aroma it imparts to the beer. This has resulted in the 'natural' selection of a group of yeast which share a common application, the industrial production of beer. The process to which these yeast are subjected has undergone little fundamental change since its inception. However, increasingly, modern commercial pressures demand the reliable production of low cost, high quality beer, requiring improvements in process control, yeast handling procedures and the yeast strains themselves. The purpose of this brief review is to outline some of the genetic strategies which have been applied to the development of new improved strains of yeast. Such strategies include mutation, hybridization and selection which have been reviewed previously (Kielland-Brandt, 1981; Freeman & Peberdy, 1983; Panchal *et al*., 1984), consequently here I will concentrate on the use of recombinant DNA technology. The following contemporary reviews on this subject should be noted (Tubb & Hammond, 1987; Hinchliffe & Vakeria, 1989; Johnston, 1990).

Yeast strain improvement affords an opportunity to improve the fermentation and post-fermentation stages of beer production, resulting in significant process and quality advantages (Table 8.1). It is apparent from the published work in the field that process costs can be reduced by providing yeast capable of secreting enzymes that would otherwise be added at the beginning or end of fermentation. Process efficiencies can be achieved by increasing volume throughput, either by minimizing the residence time of beer in vessels, or reducing beer viscosity and thereby facilitating beer filtration. Product quality may be enhanced by removing or inactivating genes responsible for the production of off-flavours or reducing the susceptibility of the fermenting beer to microbial contamination. New raw materials may be used by extending the ability of yeast to ferment novel carbon sources, and the value of spent (waste) yeast can be enhanced by introducing the genetic capability for the yeast to produce protein products of higher value than the yeast itself (Table 8.1).

Table 8.1. Strain improvement of brewing yeast

Improvement	Reference
Amylolysis – low carbohydrate beer	Vakeria & Hinchliffe (1989), Lancashire *et al.* (1989).
Proteolytic yeast – aid colloidal stability	Young & Hosford (1987)
Catabolite derepression – improved fermentation	Stewart *et al.* (1985)
Reduced diacetyl production – reduce fermenter residence time	Sone *et al.* (1987), Dillemans *et al.* (1987), Suikho *et al.* (1989), Gjermansen *et al.* (1988)
Glucanolysis – reduce beer viscosity (filtration)	Enari *et al.* (1987), Cantwell *et al.* (1987), Lancashire & Wilde (1987)
Flocculation control – beer clarification	Watari *et al.* (1987)
Reduced phenolic off-flavours	Tubb (1987)
Zymocin production – protection against contamination	Hammond & Eckersley (1984)
Enhanced value of spent yeast	Hinchliffe *et al.* (1987)
Genetic tagging – strain identification	Lancashire & Hadfield (1986)

In addition to the foregoing, molecular genetics can be used to discriminate between yeasts with similar physiological properties, which have otherwise proven difficult to differentiate (Meaden, 1990). This is an area which is receiving increasing attention, particularly with the development of internationally branded beers, brewed under license, where the licensee is often faced with the problem of ensuring that he is indeed using the specified strain of yeast. Within this context genetic tagging of yeast, to facilitate their recognition, has been suggested (Lancashire & Hadfield, 1986).

Process and regulatory constraints

It is axiomatic, if new yeast strains are being constructed to effect process efficiencies or reduce costs, that they are intended to be used for the production of existing products. Consequently, a major object of any strain improvement programme is not only the introduction of new or modified genetic characteristics, but also the retention and maintenance of existing characteristics. This means that the fermentation and flavour characteristics of the yeast must be maintained to ensure product consistency. Whilst this may appear technically trivial it is important to consider the nature of beer itself.

Beer is very different in both character and use to most biotechnological products produced by microbial fermentation. It is not a simply defined chemical entity, like an antibiotic or recombinant protein, but rather it is a complex mixture of flavour and aroma active compounds, which are present in subtly balanced proportions to give the product its unique and distinctive character. Whilst there is an ever growing body of information concerning the interaction between the key impact flavours in beer (Siebert, 1988), this remains a poorly understood area of science, principally due to the difficulties of objective flavour assessment and the limitations of analytical chemistry. Consequently, it is important that genetic manipulation should not affect the significant contribution that the yeast makes to the character of the beer, unless of course this is a specific objective of the modification.

In addition to the above, when introducing 'novel' genes into a chosen yeast strain it is important to ensure an adequate level of gene expression, and, if appropriate, protein secretion. Although these aspects are both strain and process specific, it should be remembered that, for example, the introduction of a new gene to direct the secretion of a particular enzyme, pre-supposes that the most appropriate enzyme and thus gene was chosen for the purpose. Early work in the construction of β-glucanolytic yeast to enhance beer filtration, concentrated on an endo-1,3-1,4-β-glucanase from *Bacillus subtilis* (Hinchliffe & Box, 1985; Cantwell *et al.*, 1985). This enzyme is effective at hydrolysing barley βglucan and hence reducing the viscosity of beer (Hinchliffe & Box, 1985). However, the optimum activity of the enzyme occurs at pH 6·1, whereas the enzyme is less than 20% efficient at the pH (4·2) which is normal in beer. This demands a higher level of gene expression and secretion than would otherwise be necessary if an enzyme with a more appropriate pH-activity profile had been chosen (Hinchliffe & Vakeria, 1989). In retrospect, it would have been prudent to select a gene/enzyme with a lower pH optimum, such as *EGL1*, encoding the major endoglucanase of *Trichoderma reesei* (Penttilä *et al.*, 1986; see chapter 5). This gene has been introduced into brewing yeast and expressed to effect the secretion of the enzyme, with the result that beer produced with such yeast shows a significant reduction in β-glucan content and an improved filtration performance (Penttilä *et al.*, 1987; Enari *et al.*, 1987).

Brewing is a large-scale batch process in which new yeast mass is generated during each fermentation, a proportion of which (approximately one third by weight) is used to inoculate successive fermentations. A yeast strain cultured in the laboratory, progresses through successive fermentations and is subjected to between ten and fifteen large-scale batch fermentations before final disposal. In practice this means that any one yeast may undergo 60-100 cell generations before replacement from stock culture. Thus, a genetically improved strain of yeast harbouring a newly introduced gene must possess a sufficiently stable phenotype to effect a consistent performance throughout its production life. It is therefore essential that newly introduced genes are both inheritably and structurally stable.

Coupled with these process requirements it is becoming increasingly apparent that a number of regulatory constraints will be applied to the large-scale commercial introduction of genetically improved yeast, particularly those constructed using recombinant DNA techniques (Dunn, 1987). In the United Kingdom, beers to be produced by genetically engineered yeast will be classified as novel foods, and as such will be subject to the current voluntary notification scheme which provides for Ministerial approval of their commercial use. The expert committee advising the Minister of Agriculture, Fisheries and Food (MAFF) is the Advisory Committee on Novel Foods and Process (ACNFP). This committee will use the same criterion for assessing the suitability of genetically improved yeast for beer production as is applied to any other 'novel' food, namely safety. Hence, it is necessary to ensure that such beer is safe for human consumption, being comparable in quality to products produced with unmodified brewing yeast strains. At present there is no case history upon which to base judgements of safety and risk, since recombinant DNA technology is such a recent innovation. Accordingly, it has been suggested that the genetic material introduced into brewing yeast should be, ideally, derived from a food grade organism, of GRAS (= generally regarded as safe) status. Moreover, DNA introductions should be kept to a minimum, avoiding the retention of extraneous DNA sequences wherever possible (Vakeria & Hinchliffe, 1989).

The genetic properties of brewing yeast

Brewing yeast comprise a heterogeneous collection of strains which have traditionally been classified into two groups: bottom fermenting lager yeast and top fermenting ale yeast. The phylogeny and genotypic organisation of these yeast is complex and unclear, however brewing yeast have been classified as *Saccharomyces cerevisiae* (Kreger-van-Rij, 1985; Stewart & Russell, 1986).

Virtually all strains of brewing yeast are polyploid or aneuploid, ranging from diploid to heptaploid (Sakai & Takahashi, 1972; Stewart *et al.*, 1976; Lewis *et al.*, 1976; Molzahn, 1977). Recent examination of a group of six proprietary strains of Bass yeast reflects this observed variation in ploidy (Table 8.2). Although there is some similarity between lager strains

Table 8.2. Ploidy and rDNA polymorphisms in brewing yeast

Ale Yeast	μg DNA 10^{-9} cells*	Ploidy*	rDNA form
BB1	–	–	I
BB2	139	3·3	I
BB3	233	5·5	II
NCYC24O	–		I/II
Lager Yeast			
BB9	134	3·15	II
BB10.2	102·5	2·4	II
BB10.6	107·5	2·5	II
BB11	–	2·0	II

*Ploidy estimates were performed as described by Aigle *et al.* (1983). Total genomic DNA was digested with *Eco*RI, separated by gel electrophoresis and hybridized with a 2·6 kb *Hin*dIII fragment containing the 5S rDNA (Fleming, 1988); rDNA forms I and II are as defined by Pedersen (1983).

(diploid or near diploid), the two ale strains, BB2 and BB3, show markedly different ploidy, even though the strains possess very similar brewing characteristics (Fleming, 1988).

Detailed characterization of the gross molecular arrangement of the brewing yeast genome is now possible using whole chromosome separation techniques, such as orthogonal field alternating gel electrophoresis (OFAGE), coupled with DNA:DNA hybridization. This facilitates the construction of a molecular karyotype of the yeast strain and can be used for the practical purpose of strain identification or differentiation, where traditional physiological-based methods have proved unsuccessful (Pedersen, 1987).

The Carlsberg bottom fermenting lager yeast is *probably* the most extensively studied strain of commercial brewing yeast from a genetic aspect. This yeast is reticent to mate and sporulates poorly, typical of all brewing strains. Such properties make classical genetic techniques difficult to implement in these yeast. However, genetic characterization has been facilitated, in the Carlsberg yeast, by a 'single-chromosome' transfer technique, allowing transfer of whole chromosomes into genetically defined strains of *S. cerevisiae* (Kielland-Brandt *et al.*, 1989). It has been revealed that at least two versions of chromosomes III, V, X, XII and XIII are present in the Carlsberg type strain, of which only one version is capable of homologous recombination with the corresponding chromosomes in *S. cerevisiae* (Gjermansen *et al.*, 1988). It would appear that the Carlsberg yeast contains at least two genomes, one from *S. cerevisiae* and the other from *S. monacensis* (Pedersen, 1986a & b). A similar pedigree may typify the majority of bottom fermenting lager yeast.

Restriction fragment length polymorphisms have been used to examine the phylogeny of brewing yeast. These studies confirm the observation that bottom fermenting lager yeast are more homogeneous than ale yeast, and may share a common ancestral origin (Pederson, 1983, 1985, 1986b). DNA polymorphisms in the *RDN1* locus on chromosome XII reveal distinct patterns as a result of *Eco*RI digestion (Pedersen, 1985); *S. cerevisiae* produces two forms of *RDN1*, I and II, which are typically found in top fermenting ale yeast (Pedersen, 1986b). By contrast, bottom fermenting lager yeast usually possess form II only. This pattern of bands can be seen in the data presented in Table 8.2 for a group of Bass yeast; the pattern observed is entirely consistent with the observations of Pedersen (1986b).

Genetic strategies

The complex genetic character of brewing strains together with the requirement to maintain the existing characteristics of the yeast has focused recent attention on the use of recombinant DNA techniques for strain improvement. Although cross-breeding by hybridization can make a significant contribution to strain improvement (Gjermansen & Sigsgaard, 1987), the facility to introduce novel genetic characteristics, or specifically modify existing genes, affords obvious practical benefit.

Transformation

Brewing yeast transformation was first achieved with the introduction of a recombinant 2 μm plasmid carrying the *S. cerevisiae CUP1* gene, conferring resistance to copper (Henderson *et al.*, 1985). The success of this approach can be attributed to the inherent sensitivity of brewing strains to copper ions (0·1 mM) and the presence of the endogenous 2 μm plasmid of *S. cerevisiae* in these strains (Tubb, 1980). Subsequently, many other recombinant plasmids have been introduced into brewing yeast; without exception, this has been achieved through the use of dominant selectable genes of either yeast or non-yeast origin (Table 8.3) (Knowles & Tubb, 1987; Hinchliffe & Vakeria, 1989).

We have used *CUP1* very successfully in our laboratory, to effect the transformation of a wide range of strains, and find the transformants to have perfectly acceptable brewing properties (Hinchliffe & Daubney, 1986). Whilst this is the case for *CUP1*, it is important to consider the impact of transformation and choice of selectable marker on the properties of the yeast.

Table 8.3. Dominant selectable markers used to transform brewing yeast

Gene	Selection	Reference
CUP1	copper resistance	Henderson *et al.* (1985)
ILV2	sulfometuron methyl resistance	Fleming (1988)
SMR1	sulfometuron methyl resistance	Casey *et al.* (1988)
MEL1	growth on melibiose	Tubb *et al.* (1986)
KILk1	immunity to killer toxin	Bussey & Meaden (1985)
POF1	cinnamic acid resistance	Tubb (1987)
$G418^R$	G418 (geneticin) resistance	Yocum (1986)
CM^R	chloramphenicol resistance	Hadfield *et al.* (1986)
aroA	glyphosphate resistance	Kunze *et al.* (1989)

The *ILV2* gene of *S. cerevisiae* can be used as a dominant selectable marker for yeast transformation (Falco & Dumas, 1985). *ILV2* encodes an enzyme, α-acetolactate synthase, which, when expressed at high levels on a multi-copy plasmid, confers resistance to the herbicide sulfometuron methyl (SM^R). If brewing strains are transformed to SM^R with a multi-copy *ILV2* gene, they over-produce α-acetolactate, an intermediate metabolite in the biosynthetic pathway to isoleucine and valine. Unfortunately, however, α-acetolactate is readily oxidized to the vicinal diketone, diacetyl, which imparts a distinct flavour/aroma of butterscotch/toffee to beer at levels $\geq 0 \cdot 1$ ppm ($\geq 0 \cdot 0012$ mM). Hence, yeast producing higher levels of this compound are unacceptable particularly for lager beer production. The data presented in Fig. 8.1 clearly demonstrate this effect and serve to illustrate the importance of choosing an appropriate selectable marker.

Although multiple copies of *ILV2* can result in over-production of diacetyl in beer, mutant alleles have been isolated which confer SM^R at reduced gene copies by virtue of their altered specificity for the herbicide. One such allele, SMR1-410, has been used to effect brewing strain transformation using integrating vectors, without apparent detriment to beer flavour (Casey *et al.*, 1988).

Stability of recombinant plasmids

The inheritable stability of recombinant 2 μm plasmids is greater in brewing yeast than 'laboratory' yeast, when grown in non-selective media (Hinchliffe & Daubney, 1986). This has been attributed to the higher ploidy of brewing strains, since elevated ploidy does influence 2 μm stability (Hinchliffe & Vakeria, 1989; Mead *et al.*, 1986). Interestingly, 2

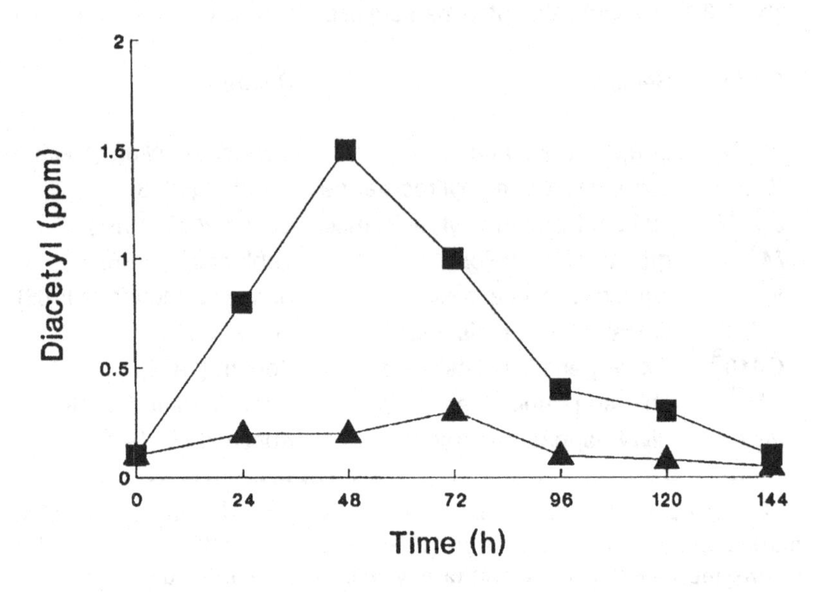

Fig. 8.1. Diacetyl production during fermentation. Brewing yeast strain BB9 was transformed to sulfometuron methyl resistance (2·5 μg ml^{-1}) with the *ILV2* plasmid pCP-2-4-10 (Falco & Dumas, 1985). BB9 (triangles) and BB (pCP-2-4-10) (squares) were subjected to fermentation in brewers' wort and samples were taken for vicinal diketone determination according to the method of Ault *et al.* (1968). Data from Fleming (1988).

μm based vectors appear to be more stable in lager yeast than ale yeast, without any obvious explanation (Meaden & Tubb, 1985; Hinchliffe & Daubney, 1986).

Many recombinant genes have now been introduced into brewing strains using 2 μm based vectors (Hinchliffe & Vakeria, 1989), and it has been suggested that 2 μm plasmids may be sufficiently stable to facilitate the large-scale use of such plasmid-bearing strains (Hinchliffe & Daubney, 1986). This may indeed prove possible, particularly in the case of strains manipulated to produce an extracellular enzyme, where, providing that there are sufficient cells in a population retaining the recombinant gene and hence secreting active enzyme, the fermentation will proceed according to expectation. Although plasmid loss will lead to a gradual reduction in the amount of enzyme supplied to the fermentation, the resultant plasmid-free cells will revert to the untransformed parental phenotype, and will not adversely influence the properties of the fermentation or resultant beer.

One complication of 2 μm-based vectors is that they exist as shuttle vectors, containing selection and maintenance functions for both yeast and *Escherichia coli*. The latter normally takes the form of an *E. coli* plasmid, such as a *Col* E1 derivative, carrying an antibiotic resistance determinant. Whilst it should be stressed that there is no evidence to suggest that such DNA sequences are either 'unsafe' or have any deleterious effect upon the properties of brewing strains (Hinchliffe & Daubney, 1986), it is desirable that these extraneous DNA sequences are not retained in the brewing strain. This issue has been addressed with the development of a new class of yeast/*E. coli* shuttle vector, designated 'disintegration vector' (Chinery & Hinchliffe, 1989; see chapter 4). These plasmids consist of a whole 2 μm plasmid together with an *E. coli* replicon and yeast selectable marker. In the case of brewing yeast transformation it is preferable to use a homologous gene such as *CUP1*, as selectable marker. The difference between disintegration vectors and other whole 2 μm plasmids, is that the bacterial replicon is inserted between two direct repeats of the 2 μm FRT (FLP recombination target site). Hence, upon yeast transformation the unwanted bacterial DNA is excised as a result of FLP mediated recombination, leaving a 2μm plasmid harbouring a selectable marker plus gene of interest (Chinery & Hinchliffe, 1989). Moreover, transformants carrying the disintegrated plasmid exhibit an extremely high degree of inheritable stability. Recent studies, albeit using cir^0 strains grown in continuous culture, reveal such plasmids to be 100% stable following 100 generations in non-selective medium (P. Stanbury, personal communication). Similarly high levels of stability have also been observed following sequential batch culture of brewing yeast (Chinery & Hinchliffe, 1989).

An alternative method of avoiding the retention of bacterial replicon DNA, using a 2 μm plasmid, has been described recently (Vakeria & Hinchliffe, 1989). In this case the *DEX1* gene of *Saccharomyces diastaticus* was introduced into brewing yeast on a 2 μm plasmid, pDVK1, which was designed to facilitate the *in vitro* excision of bacterial replicon DNA prior to transformation. Thus, plasmid pDVK1 was cleaved with the endonuclease *Xba*I and a 7·34 kb-pair DNA fragment carrying yeast DNA only was gel purified and self-ligated prior to copper resistance transformation (Vakeria & Hinchliffe, 1989). Amylolytic transformants were subsequently found to contain only yeast DNA and were sufficiently stable to be recommended for large-scale commercial application (Vakeria & Hinchliffe, 1989).

Integrative transformation

Several different procedures have now been developed for the integrative transformation of brewing strains with recombinant DNA. In practice all procedures rely upon the combined use of a dominant selectable marker and DNA sequence with chromosomal homology. However, unlike 2 μm-based systems, integrative transformation requires the selectable marker to be effective at low gene dosage; hence, genes such as *CUP1* and *ILV2* are inappropriate.

The simplest approach to integrative transformation is to use a dominant gene as the target sequence for homologous recombination into the chromosome. Sulfometuron methyl resistant alleles of *ILV2* (Casey *et al.*, 1988) have been used to direct the integration of the amyloglucosidase gene of *Schwanniomyces occidentalis* into the 3' region of the *ILV2* gene of brewing yeast (Lancashire *et al.*, 1989). Inevitably, this approach has limited application due to the availability of homologous selectable markers.

Co-transformation of linear integrating DNA with a selectable 'helper' plasmid, has been employed to integrate an endoglucanase gene from *Trichoderma reesei* (Penttilä *et al.*, 1987). In this case the β-glucanase gene, *egl1*, was fused to the phosphoglycerate kinase (*PGK*) expression signals on a *LEU2* integrating vector, pMN20 (Penttilä *et al.*, 1987). Brewing yeast were co-transformed to copper resistance with the *CUP1*-2 μm plasmid, pET13:1, and pMN20 linearized by digestion at the unique *Bst*EII site in the *LEU2* gene. Endoglucanase secreting transformants were obtained which subsequently lost the copper resistance phenotype due to segregation of pET13:1. Interestingly, copper sensitive β-glucanase producing (*Egl*$^+$)transformants were unstable for the *Egl*$^+$ phenotype and several rounds of colony purification were necessary before stable *Egl*$^+$ copper-sensitive integrants were obtained. Examination of genomic DNA revealed that integrants possessed a variable copy number (2-5 copies), resulting from integration by tandem array. This variable copy number was, presumably, responsible for the variation in β-glucanase production amongst integrants (Penttilä *et al.*, 1987). Nevertheless, integrants were capable of both hydrolysing β-glucan and reducing the viscosity of beer (Enari *et al.*, 1987).

One of the most elegant approaches to gene integration in brewing yeast has been described by Yocum (1986). Using a two step procedure, the amyloglucosidase gene (AMG) of *Aspergillus niger* was integrated, whilst avoiding retention of bacterial replicon DNA. The AMG gene was inserted into an integrating (non-replicating) plasmid within the homothallism gene (*HO*) target locus. This integrating plasmid also carried a dominant selectable marker, G418^R, and the *LacZ* gene of *E. coli* expressing β-galactosidase from a yeast promoter. As a first step, the integrating plasmid was cleaved at a unique restriction site within the *HO* target locus, and the linear DNA was then used to transform brewing yeast to G418^R.

The resultant transformants were phenotypically $G418^R$ possessing a blue colony morphology on β-galactosidase detection medium. In a second step, phenotypically white (β-galactosidase non-producing) and $G418^S$ colonies were isolated, resulting from a recombinational 'loop-out' between the duplicated *HO* target sequences. This second recombination event led to the loss of the extraneous DNA carried on the original integrating vector, the stable retention of the AMG gene and hence the production of extracellular active enzyme (Yocum, 1986).

More recently a similar two step integration procedure has been described to integrate the α-acetolactate decarboxylase (ALDC) gene of *Enterobacter aerogenes* in brewing yeast (Fujii *et al*., 1990). These authors used the $G418^R$ selection and β-galactosidase screening procedure described by Yocum (1986), but chose to integrate the ALDC gene within a homologous recombination sequence derived from either the *URA3* or the ribosomal DNA (rDNA). In both cases integration was directed by linearizing the vector within the target sequence prior to transformation to $G418^R$. Interestingly, targeting recombination to the rDNA locus resulted in a twenty-fold higher frequency of transformation than targeting at the *URA3* locus. When integrants were analyzed, all transformants integrated at the rDNA locus showed about 30 to 50-fold higher ALDC activity than those integrated at *URA3*. It was assumed that this difference was due to multiple copies of the plasmid integrated at the rDNA locus. Furthermore, it was shown that brewing yeast clones with increased ALDC activity could be isolated by repeated transformation and marker excision of the ALDC positive transformants (Fujii *et al*., 1990).

Examination of the ALDC producing transformants revealed that the marker-excised plasmid, that is, integrants containing the ALDC gene and no extraneous DNA, were considerably more stable under conditions of non-selective growth than corresponding integrants containing both marker-excised and the complete plasmid. This was attributed to a higher frequency of spontaneous homologous recombination between the plasmid and chromosomal rDNA sequences, which does not occur to the same extent in marker-excised clones. Fermentation experiments with ALDC producing integrants indicated that the yeast produced sufficient enzyme to reduce significantly the total diacetyl in the fermented wort. Although, interestingly, integrants did not produce as much enzyme as 2 μm-based plasmid clones (Fujii *et al*., 1990).

A further method of gene integration, designed specifically for genetically tagging brewing yeast has been described (Lancashire & Hadfield, 1986). Here, a dominant selectable marker conferring chloramphenicol resistance was inserted into the chromosomal *HIS3* gene *in vitro*. This sequence of DNA was then linearized, producing a *HIS3* gene with 5' and

3' termini which was used to transform yeast to chloramphenicol resistance. Integration into the chromosomal *HIS3* target sequence was confirmed by Southern hybridization and a second transformation undertaken. In the case of the latter the chloramphenicol resistance gene contained within the target *HIS3* sequence was replaced with a 15 base-pair 'tag' sequence. Chloramphenicol resistant transformants were transformed with this new 'tag' sequence, and cells were plated onto non-selective medium and screened for chloramphenicol sensitivity. Sensitive derivatives, in which the chloramphenicol resistance gene had been replaced by the 'tag' sequence, were then verified by a combination of hybridization procedures (Lancashire & Hadfield, 1986). Clearly, this procedure could also be used to integrate genes of interest, other than genetic tags, providing that integration did not affect either the strain or expression of the integrated gene. One might reasonably contemplate using this method for *in vitro* mutagenesis of brewing yeast genes.

In vitro *mutagenesis*

In vitro mutagenesis affords extensive opportunities for specifically inactivating yeast genes to modify the phenotypic properties of the strain. However, to apply such techniques successfully it is necessary to have a good understanding of the biochemistry and genetics of the particular phenotype to be modified. The reduction of diacetyl formation by brewers yeast is ideally placed to benefit from this kind of technique.

The vicinal diketones, diacetyl and 2,3-pentanedione are normal products of yeast metabolism and are formed from α-acetolactate and α-aceto-α-hydroxybutyrate respectively. These two products are the first intermediates in the common pathways for the biosynthesis of isoleucine and valine and the products of the enzymic action of α-acetolactate synthase on pyruvate and α-ketobutyrate. α-Acetolactate synthase is the product of the *ILV2* gene of yeast, which is present as two different alleles in the Carlsberg type yeast, one of which is homologous with the *S. cerevisiae ILV2* gene, the other with that from *S. carlsbergensis* (Gjermansen *et al.*, 1988). In order to construct *ilv2* mutants in the Carlsberg type yeast, Gjermansen *et al.* (1988) constructed *in vitro* deletions in both alleles of the *ILV2* gene. In the case of the *S. cerevisiae* homologue a simple internal deletion was created and the resultant gene incorporated into a $G418^R$ integration 'loop-out' vector analogous to that described by Yocum (1986), with the exception that the target homologous *HO* locus had been replaced by the *ILV2* homologous sequence. Thus, brewing yeast were transformed to $G418^R$ and integrants were subsequently isolated in which the $G418^R$ determinant and corresponding *lacZ* (β-galactosidase producing) gene had been excised (Gjermansen *et al.*, 1988). By contrast, the other *ILV2* allele, with *S. carlsbergensis* homology, received both an *in vitro* deletion and an insertion of the cloned *ILV5* gene. This construct was similarly incorporated into a G418^R integration 'loop-out' vector with a view to effecting brewing yeast transformation (Gjermansen *et al.*, 1988). The purpose of the *ILV5* gene insertion was to increase the copy number of this gene and hence elevate the level of expression of the corresponding reductoisomerase, which had previously been shown to reduce the propensity of yeast to produce diacetyl (Dillemans *et al.*, 1987).

It remains to be seen whether yeast manipulated using *in vitro* mutagenic techniques do produce reduced quantities of the flavour active vicinal diketone, diacetyl. However, the work which has been performed thus far, certainly suggests that these yeast will prove suitable for commercial implementation. Moreover, unlike yeasts which express α-acetolactate decarboxylase from the *Enterobacter aerogenes* gene (Fujii *et al.*, 1990), the strain constructed by Gjermansen *et al.* (1988) could be truly classified as 'self-cloned', containing yeast DNA in yeast.

Conclusions

Brewing yeast can be simply classified as those yeast capable of producing palatable beer. They comprise a heterogeneous collection of strains belonging to the genus *Saccharomyces cerevisiae*, each strain imparting its own distinctive flavour to the fermented beer. Historically, these yeast have undergone little or no improvement for their particular application, even though their contribution to the character and quality of beer is so important. However, with the advent of molecular genetic technology it is now possible to construct new and improved strains for commercial application. Here I have briefly described some of the genetic strategies which have been developed to effect the genetic modification of brewing yeast with recombinant DNA. It is apparent from the work in this area that most of the more obvious targets for brewing yeast strain improvement will be resolved within the next two or three years. The challenge for molecular biologists will then be to combine each of the individual improvements within their chosen yeast strains. Thus, it will be necessary to combine novel combinations of genes to create strains which are, for example, both capable of fermenting dextrins and at the same time hydrolyzing β-glucan present in beer. It will be interesting to see whether this will create undue physiological stress on the yeast, perturbing their innate brewing properties, and thereby creating new technical challenges to resolve.

In addition to the outstanding technical issues relating to the construction of new improved strains of brewing yeast, it will be necessary for the industry to consider regulatory approval for the use of genetically modified brewing yeast. The genetic strategies which have been developed leave me in little doubt that the approval process will be straightforward. However, the implementation of genetically modified yeast for the production of commercial beers may be complicated by the risk of adverse consumer reaction. The brewing industry is naturally cautious and traditionally in the U.K. very close to the market. Thus, if there is the possibility of adverse consumer reaction associated with the public disclosure of genetically modified organisms, this could significantly reduce the perceived benefit derived by the industry from the genetic improvement. Hence, it is essential that the method by which yeast are genetically modified and the associated benefits are clearly and responsibly communicated to the consumer. Such considerations may influence the rate of implementation of this technology in breweries worldwide.

Acknowledgements I would like to thank the Directors of Bass Brewers Limited for permission to publish this paper.

References

Aigle, M., Erbs, D. & Moll, M. (1983). Determination of brewing yeast ploidy by DNA measurement. *Journal of the Institute of Brewing*, **89**, 72-74.

Ault, R. G., Jones, M. O., Hudson, J. R., Howard, G. A., Martin, P. A., Bayles, P. F., Bishop, L. R., Greenshields, R. N., Taylor, L. & Sinclair, A. (1968). The Institute of Brewing Analysis Committee. The determination of vicinal diketones in beer. *Journal of the Institute of Brewing* **74**, 196-199.

Bussey, H. & Meaden, P. (1985). Selection and stability of yeast transformants expressing cDNA of an M1 killer toxin-immunity gene. *Current Genetics* **9**, 285-291.

Cantwell, B. A., Brazil, G., Hurley, J. & McConnell, D. (1985). Expression of the gene for the endo-β-1,3-1,4-glucanase from *Bacillus subtilis* in *Saccharomyces cerevisiae*. *Proceedings of the European Brewery Convention, 20th Congress, Helsinki*, 259-266.

Cantwell, B. A., Ryan, T., Hurley, J. C., Doherty, M., McConnell, D. J. (1987). Degradation of barley β-glucan by brewers' yeast. *European Brewery Convention, Symposium on Brewers' Yeast, Vuoranta, (Helsinki), Monograph XII*, 186-196.

Casey, G. P., Xiao, W., Rank, G. H. (1988). A convenient dominant selection marker for gene transfer in industrial strains of *Saccharomyces* yeast, *SMIR* encoded resistance to the herbicide sulfometuron methyl. *Journal of the Institute of Brewing* **94**, 93-97.

Chinery, S. A. & Hinchliffe, E. (1989). A novel class of vector for yeast transformation. *Current Genetics* **16**, 21-25.

Dillemans, M., Goosens, E., Goffin, O. & Masschelein, C. A. (1987). The amplification effect of the *ILV5* gene on the production of vicinal diketones in *Saccharomyces cerevisiae*. *Journal of the American Society of Brewing Chemists* **45**, 81-84.

Dunn, A. J. (1987). Food safety approval for genetically modified yeast strains. *European Brewery Convention Symposium on Brewers' Yeast, Vuoranta, (Helsinki), Monograph XII*, 223-233.

Enari, T. -M., Knowles, J., Lehtinen, U., Nikkola, M., Penttilä, M., Suikho, M. -L., Home, S. & Vilpola, A. (1987). Glucanolytic brewers' yeast. *Proceedings of the European Brewery Convention, 21st Congress, Madrid*, 529-536.

Falco, S. C. & Dumas, K. S. (1985). Genetic analysis of mutants of *Saccharomyces cerevisiae* resistant to the herbicide sulfometuron methyl. *Genetics* **109**, 21-35.

Fleming, C. J. (1988). The genetic manipulation of brewing yeasts: the inheritance of 2 μm plasmids. Ph.D. Thesis, CNAA.

Freeman, R. F. & Peberdy, J. F. (1983). Protoplast fusion in yeasts. In *Yeast Genetics Fundamental and Applied Aspects*, (ed. J. F. T. Spencer, D. M. Spencer & A. R. W. Smith), pp. 243-253. Springer-Verlag: New York.

Fujii, T., Kondo, K., Shimizu, F., Sone, H., Tanaka, J. -I. & Inoue, T. (1990). Application of a ribosomal DNA integration vector in the construction of a brewers' yeast having α-acetolactate decarboxylase activity. *Applied and Environmental Microbiology* **56**, 997-1003.

Gjermansen, C. & Sigsgaard, P. (1987). Cross-breeding with brewers' yeast and evaluation of strains. *European Brewery Convention Symposium on Brewers' Yeast, Vuoranta, (Helsinki), Monograph XII*, 156-168.

Gjermansen, C., Nilsson-Tilgren, T., Petersen, J. G. L., Kielland-Brandt, M. C., Sigsgaard, P. & Holmberg, S. (1988). Towards diacetyl-less brewers' yeast. Influence of *ilv2* and *ilv5* mutations. *Journal of Basic Microbiology* **28**, 175-183.

Hadfield, C., Cashmore, A. M. & Meacock, P. A. (1986). An efficient chloramphenicol-resistance marker for *Saccharomyces cerevisiae* and *Escherichia coli*. *Gene* **45**, 149-158.

Hammond, J. R. M. & Eckersley, K. W. (1984). Fermentation properties of brewing yeast with killer character. *Journal of the Institute of Brewing* **90**, 167-177.

Henderson, R. C. A., Cox, B. S. & Tubb, R. S. (1985). The transformation of brewing yeast with a plasmid containing the gene for copper resistance. *Current Genetics* **9**, 133-138.

Hinchliffe, E. & Box, W. G. (1985). Beer enzymes and genes: the application of a concerted approach to β-glucan degradation. *Proceedings of the European Brewery Convention, 20th Congress, Helsinki*, 267-274.

Hinchliffe, E. & Daubney, C. J. (1986). The genetic modification of brewing yeast with recombinant DNA. *Journal of the American Society of Brewing Chemists* **44**, 98-101.

Hinchliffe, E., Kenny, E. & Leaker, A. (1987). Novel products from surplus yeast via recombinant DNA technology. *European Brewery Convention Symposium on Brewers' Yeast, Vuoranta (Helsinki), Monograph XII*, 139-154.

Hinchliffe, E. & Vakeria, D. (1989). Genetic manipulation of brewing yeasts. In *Molecular and Cell Biology of Yeasts*, (ed. E. F. Walton & G. T. Yarranton), pp 280-303. Blackie: Glasgow and London.

Johnston, J. R. (1990). Brewing and distilling yeasts. In *Yeast Technology*, ed. J. F. T. Spencer, & D. M. Spencer, pp. 55-104. Springer-Verlag: New York.

Kielland-Brandt, M. C. (1981). The genetics of brewers' yeast. *Proceedings of the European Brewery Convention, 18th Congress, Copenhagen*, 263-276.

Kielland-Brandt, M. C., Gjermansen, C., Nilsson-Tilgren, T. & Holmberg, S. (1989). Yeast breeding. *Proceedings of the European Brewery Convention, 22nd congress, Zurich*, 37-47.

Knowles, J. K. C. & Tubb, R. S. (1987). Recombinant DNA: gene transfer and expression techniques with industrial yeast strains. *European Brewery Convention Symposium on Brewers' Yeast, Vuoranta, (Helsinki), Monograph XII*, 169-185.

Kreger-van-Rij, N. J. W. (1984). *The Yeasts: A Taxonomic Study*, 3rd edn. Elsevier: Amsterdam.

Kunze, G., Bode, R., Rintala, H. & Hofmeister, J. (1989). Heterologous gene expression of the glyphosphate resistance marker and its application in yeast transformation. *Current Genetics* **15**, 91-98.

Lancashire, W. E. & Hadfield, C. (1986). Tagging of Micro-organisms. European Patent Application No. 86309272.2.

Lancashire, W. E. & Wilde, R. J. (1987). Secretion of foreign proteins by brewing yeast. *Proceedings of the European Brewery Convention, 21st Congress, Madrid*, 513-520.

Lancashire, W. E., Carter, A. T., Howard, J. J. & Wilde, R. J. (1989). Superattenuating brewing yeast. *Proceedings of the European Brewery Convention, 22nd Congress, Zurich*, 491-498.

Lewis, C. W., Johnston, J. R. & Martin, P. A. (1976). The genetics of yeast flocculation. *Journal of the Institute of Brewing* **82**, 158-160.

Mead, D. J., Gardner, D. C. J. & Oliver, S. G. (1986). Enhanced stability of 2 μm-based recombinant plasmid in diploid yeast. *Biotechnology Letters* **8**, 391-396.

Meaden, P. G. (1990). DNA fingerprinting of brewers' yeast: current perspectives. *Journal of the Institute of Brewing* **96**, 195-200.

Meaden, P. G. & Tubb, R. S. (1985). A plasmid vector system for the genetic manipulation of brewing strains. *Proceedings of the European Brewery Convention, 20th Congress, Helsinki*, 219-226.

Molzahn, S. W. (1977). A new approach to the application of genetics to brewing yeast. *Journal of the American Society of Brewing Chemists* **35**, 54-59.

Panchal, C. J., Russell, I., Sills, A. M. & Stewart, G. G. (1984). Genetic manipulation of brewing and related yeast strains. *Food Technology* **38**, 99-106.

Pedersen, M. B. (1983). DNA sequence polymorphisms in the genus *Saccharomyces*. I. Comparison of the *HIS4* and ribosomal RNA genes in lager strains, ale strains and various species. *Carlsberg Research Communications* **48**, 485-503.

Pedersen, M. B. (1985). DNA sequence polymorphisms in the genus *Saccharomyces*. II. Analysis of the genes *RDN1*, *HIS4*, *LEU2*, and *Ty* transposable elements in Carlsberg, Tuborg and 22 Bavarian brewing strains. *Carlsberg Research Communications* **50**, 263-272.

Pedersen, M. B. (1986a). DNA sequence polymorphisms in the genus *Saccharomyces*. III. Restriction endonuclease fragment patterns of chromosomal regions in brewing and other strains. *Carlsberg Research Communications* **51**, 163-183.

Pedersen, M. B. (1986b). DNA sequence polymorphisms in the genus *Saccharomyces*. IV. Homoeologous chromosomes III of *Saccharomyces bayanus*, *S. carlsbergensis* and *S. uvarum*. *Carlsberg Research Communications* **51**, 185-202.

Pedersen, M. B. (1987). Practical use of electro-karyotypes for brewing yeast identification. *Proceedings of the European Brewery Convention, 21st Congress, Madrid*, 489-496.

Penttilä, M., Lehtovaara, P., Nevalainen, H., Bhikhabhai, R. & Knowles, J. K. C. (1986). Homology between cellulase genes of *Trichoderma reesei*: complete nucleotide sequence of the endoglucanase I gene. *Gene* **45**, 253-263.

Penttilä, M., Süihko, M. -L., Lehtinen, U., Nikkola, M. & Knowles, J. K. C. (1987). Construction of brewers' yeasts secreting fungal endo-β-glucanase. *Current Genetics* **12**, 413-420.

Sakai, K. & Takahashi, T. (1972). Estimation of ploidies in brewing yeast. *Bulletin of Brewing Science* **18**, 29-36.

Siebert, K. J. (1988). A data base management system for flavour threshold information and an evaluation of strategies for identifying new flavour-active substances in beer. *Journal of the American Society of Brewing Chemists* **46**, 82-91.

Sone, H., Kondo, K., Fujii, T., Shimuzu, F., Tanaka, J. & Inoue, T. (1987). Fermentation properties of brewers' yeast having α-acetolactate decarboxylase gene. *Proceedings of the European Brewery Convention, 21st Congress, Madrid*, 545-552.

Stewart, G. G., Russell, I. & Goring, T. (1976). Nature-nurture anomalies. Further studies in yeast. *Journal of the American Society of Brewing Chemists* **33**, 137-147.

Stewart, G. G., Jones, R. & Russell, I. (1985). The use of derepressed yeast mutants in the fermentation of brewery worts. *Proceedings of the European Brewery Convention, 20th Congress, Helsinki*, 243-250.

Stewart, G. G. & Russell, I. (1986). One hundred years of yeast research and development in the brewing industry. *Journal of the Institute of Brewing* **92**, 537-538.

Suikho, M. -L., Penttilä, M., Sone, H., Home, S., Blomquist, K., Tanaka, J., Inoue, T. & Knowles, J. K. C. (1989). Pilot-brewing with α-acetolactate decarboxylase active yeasts. *Proceedings of the European Brewery Convention, 22nd Congress, Zurich*, 483-490.

Tubb, R. S. (1980). 2 μm DNA plasmid in brewery yeasts. *Journal of the Institute of Brewing* **86**, 78-80.

Tubb, R. S. (1987). Gene technology for industrial yeasts. *Journal of the Institute of Brewing* **93**, 91-96.

Tubb, R. S. & Hammond, J. R. M. (1987). Yeast genetics. In *Brewing Microbiology*, ed. F. G. Priest & I. Campbell, pp. 67-82. Elsevier Applied Sciences: Amsterdam.

Tubb, R. S., Liljestrom, P. L., Torkkeli, T. & Korhola, M. (1986). Melibiase (MEL) genes in brewing and distilling yeasts. In *Proceedings of the Institute of Brewing, Aviemore Symposium*, (ed. F. G. Priest & I. Campbell), pp. 298-303. Aberdeen University Press.

Vakeria, D. & Hinchliffe, E. (1989). Amylolytic brewing yeast: their commercial and legislative acceptability. *Proceedings of the European Brewery Convention, 22nd Congress, Zurich*, 475-482.

Watari, J., Takata, Y., Nishikawa, N. & Kamada, K. (1987). Cloning of a gene controlling yeast flocculence. *Proceedings of the European Brewery Convention, 21st Congress, Madrid*, 537-544.

Yocum, R. R. (1986). Genetic engineering of industrial yeasts. *Proceedings of Bio/Expo* 86, pp. 171-180. Butterworth: Stoneham.

Young, T. W. & Hosford, E. A. (1987). Genetic manipulation of *Saccharomyces cerevisiae* to produce extracellular protease. *Proceedings of the European Brewery Convention, 21st Congress, Madrid*, 521-528.

Chapter 9

Identification of the *Cephalosporium acremonium* *pcb*AB gene using predictions from an evolutionary hypothesis

Paul L. Skatrud, JoAnn Hoskins, Matthew B. Tobin, James R. Miller, John S. Wood, Steven Kovacevic & Stephen W. Queener

The sulphur-containing β-lactam antibiotics (penicillins, cephalosporins and cephamycins) are an extremely important and diverse group of clinically active compounds which share the four-membered β-lactam ring structure (Fig. 9.1). Due to this structural similarity, all of these compounds kill bacteria by the same process – inhibition of bacterial transpeptidases involved in the formation of peptidoglycan, an essential component of the bacterial cell wall. These β-lactam antibiotics possess a high therapeutic index as compared to many other clinically utilized antibiotics because the targeted activity is not found in mammalian cells and β-lactams have extremely potent antibacterial activity.

A wide variety of microorganisms make sulphur-containing β-lactam antibiotics, including both eukaryotes and prokaryotes. The filamentous fungi *Penicillium chrysogenum* and *Aspergillus nidulans* produce penicillin G. Another filamentous fungus, *Cephalosporium acremonium*, produces cephalosporin C. Cephamycin C is made by prokaryotes, including *Streptomyces clavuligerus*. The biosynthetic pathways which lead to penicillin G, cephalosporin C, and cephamycin C have been elucidated (Fig. 9.1) (Queener & Neuss, 1982; Jensen, 1985). The first enzymatic steps in the synthesis of β-lactams are common to all these pathways, while later steps are specific for either penicillins, cephalosporin C, or cephamycin C. The nomenclature describing the genes encoding these β-lactam synthesizing enzymes reflects the common intermediates in these pathways which allows them to be depicted in the format of a single branched pathway (*pcb* = penicillin and cephalosporin biosynthesis, *cef* = specific for cephalosporin biosynthesis, and *pen* = specific for penicillin biosynthesis; Ingolia & Queener, 1989). In order to accommodate fully the rationale used to establish this nomenclature, we recommend that those genes specific for cephamycin C biosynthesis (previously *cef*H, *cef*I and *cef*J) be referred to as *cmc* genes (i.e. *cmc*H, *cmc*I and *cmc*J) as suggested by Chen

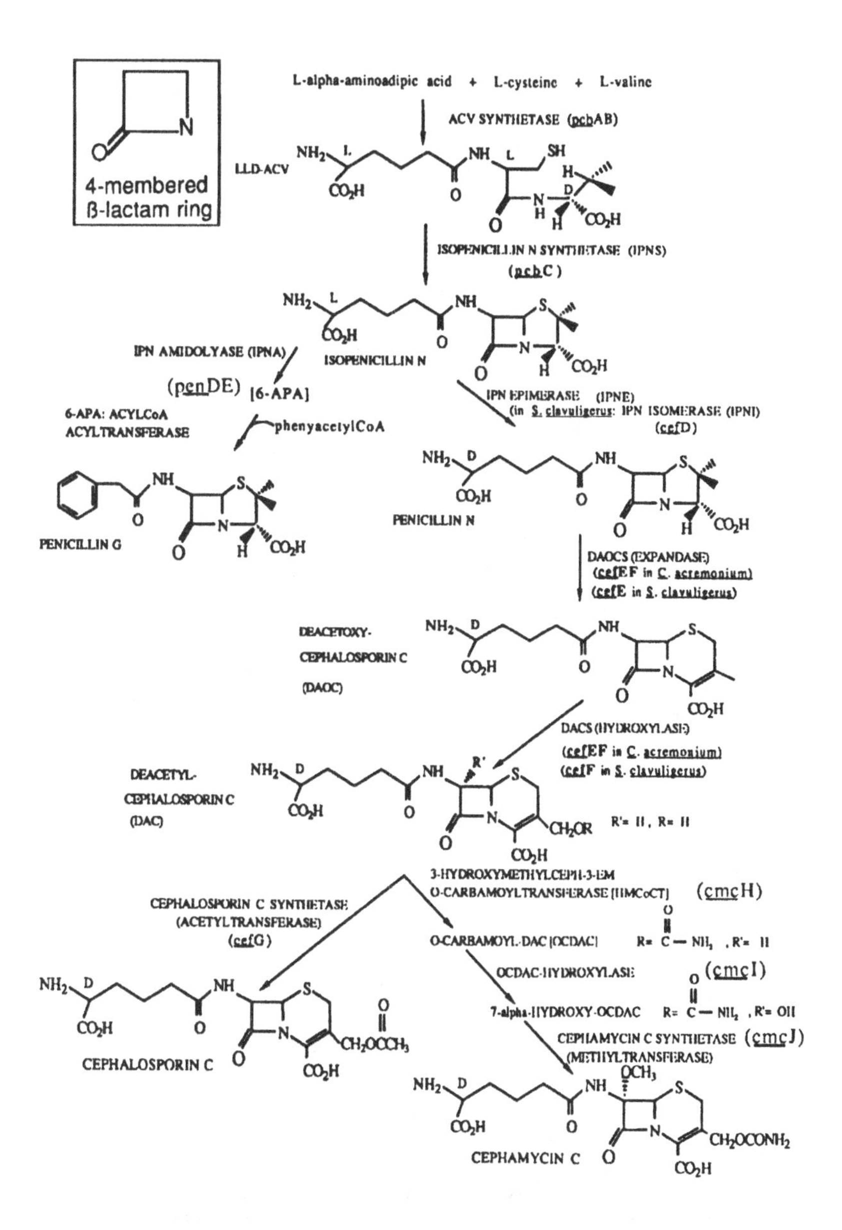

Fig. 9.1. β-lactam biosynthetic pathways to penicillin G, cephalospoin C and cephamycin.

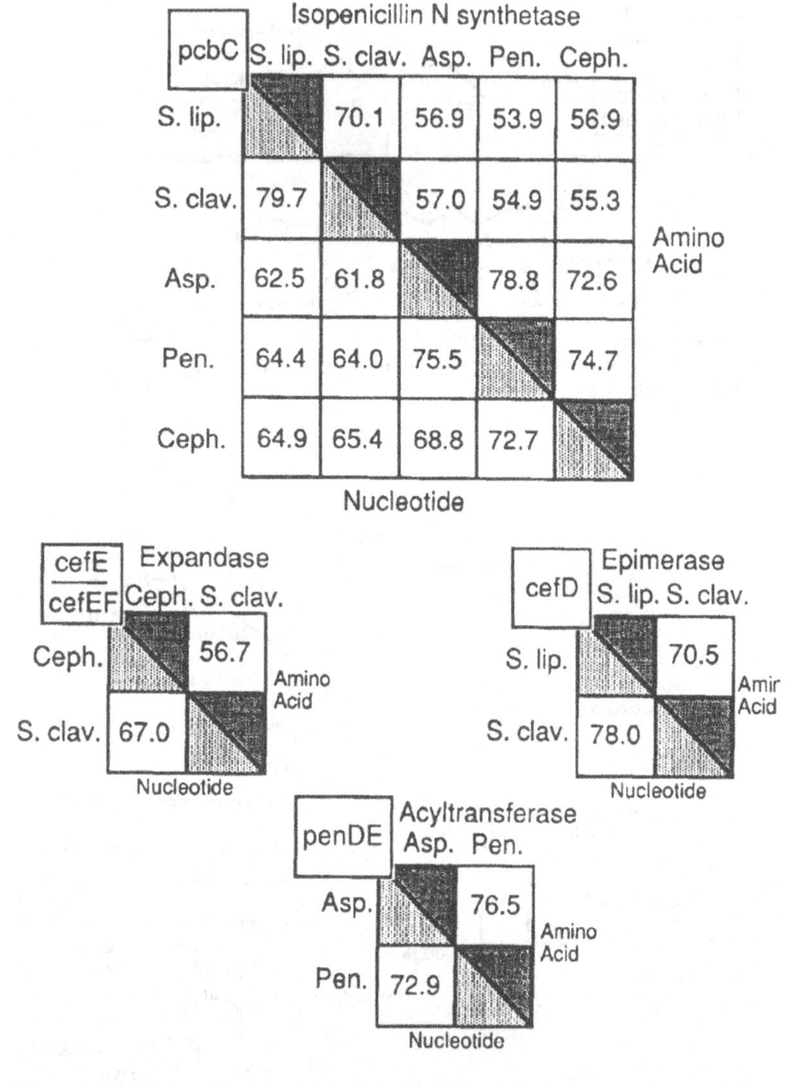

Fig. 9.2. DNA sequence (lower left) and amino acid sequence (upper right) similarities of cloned β-lactam biosynthetic genes. Numbers indicate the percent similarity for a particular gene or protein in the organisms being compared. S. clav. = *Streptomyces clavuligerus*, S. lip. = *Streptomyces lipmanii*, Asp. = *Aspergillus nidulans*, Pen. = *Penicillium chrysogenum* and Ceph. = *Cephalosporium acremonium*. (Adapted from Ingolia & Queener, 1989).

et al. (1988). This revised nomenclature appropriately recognizes those genes that function only in cephamycin C biosynthesis.

The application of recombinant DNA technology (Queener *et al.*, 1985; Skatrud *et al.*, 1986, 1987, 1989b; Cantwell *et al.*, 1990) to the industrially utilized β-lactam producing fungi has provided the basis for an hypothesis concerning the evolution of sulphur-containing β-lactam biosynthetic genes (Carr *et al.*, 1986; Weigel *et al.*, 1988). This evolutionary hypothesis predicts certain juxtapositions of these β-lactam biosynthetic genes, clustering of these genes in the filamentous fungi, and perhaps an example of gene fusion for the origin of one of the enzymes in the pathway.

Evolution of the β-lactam biosynthetic pathway

Several genes from the filamentous fungi involved in β-lactam biosynthesis have been cloned and studied at the molecular level (*pcb*C: Samson *et al.*, 1985; Carr *et al.*, 1986; Ramon *et al.*, 1987; Weigel *et al.*, 1988; *cef*EF: Samson *et al.*, 1987; and *pen*DE: Veenstra *et al.*, 1989; Tobin *et al.*, 1990). In addition to the fungal genes, β-lactam biosynthetic genes have been cloned from prokaryotic sources. For example, *pcb*C (Weigel *et al.*, 1988; Shiffman *et al.*, 1988), *cef*E (Kovacevic *et al.*, 1989), *cef*D (Kovacevic *et al.*, 1990) have been characterized at the molecular level in *Streptomyces* species. Recently, the gene which encodes lysine-ε-aminotransferase (LAT) was mapped to a position between the *pcb*C and *cef*E genes in *S. clavuligerus* (Madduri, Studdard & Vining, 1990). Based on a comparison of DNA sequence information obtained from the *pcb*C genes (Fig. 9.2), an evolutionary hypothesis was first suggested by Ingolia (Carr *et al.*, 1986; Weigel *et al.*, 1988). This hypothesis proposed horizontal transfer of the β-lactam biosynthetic genes from a prokaryote to a eukaryote approximately 370 million years ago (Fig. 9.3). In addition to *pcb*C, all other genes from the β-lactam biosynthetic pathway which have been cloned and characterized at the molecular level show similar DNA and amino acid sequence conservation supporting the horizontal transfer theory (Fig. 9.2).

The genomic arrangement of the β-lactam biosynthetic genes has also been informative with respect to the evolutionary process. Recent investigations of the prokaryote *Streptomyces clavuligerus* indicate that genes involved in β-lactam biosynthesis, *pcb*AB (S. E. Jemsem & P. L. Skatrud, unpublished results), *pcb*C (Weigel *et al.*, 1988; Shiffman *et al.*, 1988), *cef*D (Kovacevic *et al.*, 1990), *cef*E (Miller & Ingolia, 1989; Kovacevic *et al.*, 1989), and the gene which encodes LAT (Madduri *et al.*, 1990) are all physically linked within approximately 20 to 30 kb of one another. Additionally, the genes involved in β-lactam synthesis in the filamentous fungi *P. chrysogenum* and *A. nidulans*, are clustered (Smith *et al.*, 1990; Tobin *et al.*, 1990). The similar physical arrangement of β-lactam biosynthetic

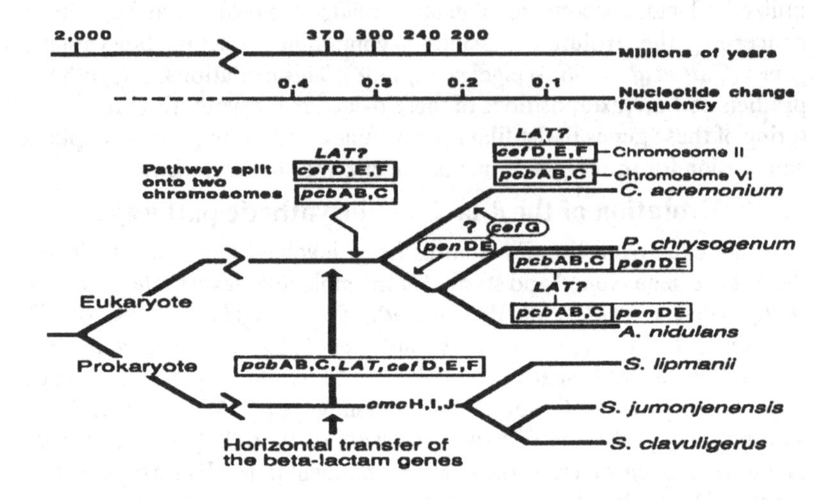

Fig. 9.3. Phylogenetic tree based on sequence identities of cloned *pcb*C genes. (Adapted from Weigel *et al.*, 1988).

genes in these two fungi may reflect their prokaryotic origin. The clustering of the biosynthetic genes also suggests that all genes required for cephalosporin biosynthesis were transferred to a eukaryotic ancestor common to *C. acremonium* and *P. chrysogenum* in a single event.

In contrast to the gene clustering in *P. chrysogenum* and *A. nidulans*, two genes (*pcb*C & *cef*EF) involved in cephalosporin C biosynthesis in *C. acremonium*, are not physically linked (Skatrud & Queener, 1989). Molecular karyotype analysis indicated that the *pcb*C gene resides on chromosome VI; the *cef*EF gene is located on chromosome II. In light of these results and considering the horizontal transfer hypothesis, it is interesting to speculate that the *pcb*C gene and other genes prior to the first branch in the biosynthetic pathway might be clustered on chromosome VI in *C. acremonium*. Furthermore, the *cef*EF gene might be associated with another cluster of genes specific for cephalosporin C biosynthesis (*cef*D and *cef*EF) on chromosome II (Fig. 9.3).

The horizontal transfer hypothesis coupled with the observed physical arrangement of genes suggests at least one explanation for the lack of cephalosporin specific genes in *P. chrysogenum* and *A. nidulans*. According to this hypothesis the β-lactam biosynthetic genes were transferred originally together in a single block from a prokaryote, perhaps an ancestor of one of the Streptomycetes, to a common ancestor of *C. acremonium*, *A. nidulans* and *P. chrysogenum*. In that common eukaryotic ancestor, the block of β-lactam biosynthetic genes was split onto two chromosomes, separating the early part of the pathway from the branch that leads to cephalosporin C synthesis. The presence of *pcb*C and *cef*EF on separate chromosomes in a modern strain of *C. acremonium* is consistent with this suggestion.

Approximately 70 million years after the horizontal transfer (see Fig. 9.3), an ancestor common to *P. chrysogenum* and *A. nidulans* diverged from the *C. acremonium* lineage. If a cluster of genes encoding *pcb*C and *pcb*AB resided on a chromosome separate from the cephalosporin-specific genes in the ancient common ancestor, then it is possible to envisage how the more recent ancestor common to *P. chrysogenum* and *A. nidulans* received only the early biosynthetic genes (*pcb*AB and *pcb*C) during the course of evolution.

The origin of the last gene (*pen*DE, Fig. 9.1) required for synthesis of penicillin G is not obvious. There is no known prokaryotic gene analogous to *pen*DE. The only β-lactam biosynthetic gene characterized to date which contains introns is the *pen*DE gene (Veenstra *et al.*, 1989). The introns in the *pen*DE gene from both *A. nidulans* and *P. chrysogenum* occur in precisely the same position (Tobin *et al.*, 1990). These data are consistent with the origin of a functional *pen*DE gene in the eukaryotic ancestor common to both *P. chrysogenum* and *A. nidulans*.

If one accepts this theory, then it is difficult to understand why the *pen*DE gene would be linked to the *pcb*C gene in the fungi, possibly an ancestral prokaryotic gene, or genes, functionally related in some manner to *pen*DE, was transferred along with *pcb*C and later modified to current functional form. The *pen*DE gene produces a multifunctional enzyme which removes the L-α-aminoadipyl side chain from isopenicillin N (isopenicillin N amidolyase) and replaces it with the CoA derivative of phenylacetic acid (acylCoA:6-APA acyltransferase) to form penicillin G. A tightly regulated gene encoding an isopenicillin N amidolyase, whose expression was sensitive to the external concentration of β-lactam, may have been present in the prokaryotic ancestor adjacent to *pcb*C. When the level of β-lactam became toxic to the producing organism, the amidolyase gene was transcriptionally activated. The amidolyase removed the L-α-aminoadipyl side chain from isopenicillin N to form 6-APA which was less toxic to the producing organism and prevented further accumulation of the toxic β-lactam. Thus the ancient amidolyase gene, if it existed, could be considered a resistance gene in its primordial host. Genetic linkage of a resistance gene to antibiotic biosynthetic genes is quite common in *Streptomyces* species (Seno & Baltz, 1989). Since the fungi are not sensitive to β-lactam antibiotics there would have been no selective advantage in having a fungal amidolyase. However, if the amidolyase was coupled to the capacity to replace the L-α-aminoadipyl side chain with a side chain creating a compound with greater antibacterial activity, then a selective advantage may have resulted. The presence of introns in *pen*DE suggests eukaryotic exon shuffling resulting in the modern day gene. All introns in the *pen*DE gene are located within the first half of the open reading frame. The first half of the *pen*DE gene is about 58% G + C in the third position of the reading frame. In contrast, the second half of the gene, which is devoid of introns, is 71% G + C in the third position of the reading frame. This shift in third position bias might suggest fusion of two genes, perhaps from different species.

Predictions based on the evolutionary hypothesis

Several testable predictions with regard to the β-lactam biosynthetic genes in the fungi have been made as a result of speculation based on the horizontal transfer hypothesis. First, genes associated with penicillin G synthesis should be clustered in fungi. This is indeed the case for *P. chrysogenum* in which all genes (*pcb*AB, *pcb*C, and *pen*DE) necessary for production of penicillin G from amino acid precursors are tightly linked (Smith *et al.*, 1990). Tobin *et al.* (1990), demonstrated that *pen*DE occupies a similar position relative to *pcb*C in *A. nidulans* as compared to *P. chrysogenum*. The *pcb*AB gene is located upstream of *pcb*C in *A. nidulans* and *P. chrysogenum* (P. L. Skatrud & J. Hoskins, unpublished results). Thus the first prediction has been fulfilled.

Next, the *pen*DE gene of *A. nidulans* and *P. chrysogenum* may be of dual origin (i.e. part prokaryote and part eukaryote). An ancestral form of the amidolyase portion of acyltransferase may yet reside in *S. clavuligerus*. If true, then a region of DNA in *S. clavuligerus* adjacent to *pcb*C may share approximately 64% sequence identity (refer to Fig. 9.2) with at least a portion of *pen*DE. Furthermore, a vestigial gene might be present in *C. acremonium*, downstream of *pcb*C which bears a predictable resemblance to at least a portion of *pen*DE and the proposed ancestral amidolyase in *S. clavuligerus*. These predictions can be examined as DNA sequence information becomes available for these regions.

The *cef*G gene is probably of prokaryotic origin. *Streptomyces lipmannii* produces a 7-α-methoxycephalosporin C; this streptomycete mat possess a gene analogous to *cef*G from *C. acremonium*. The horizontal transfer

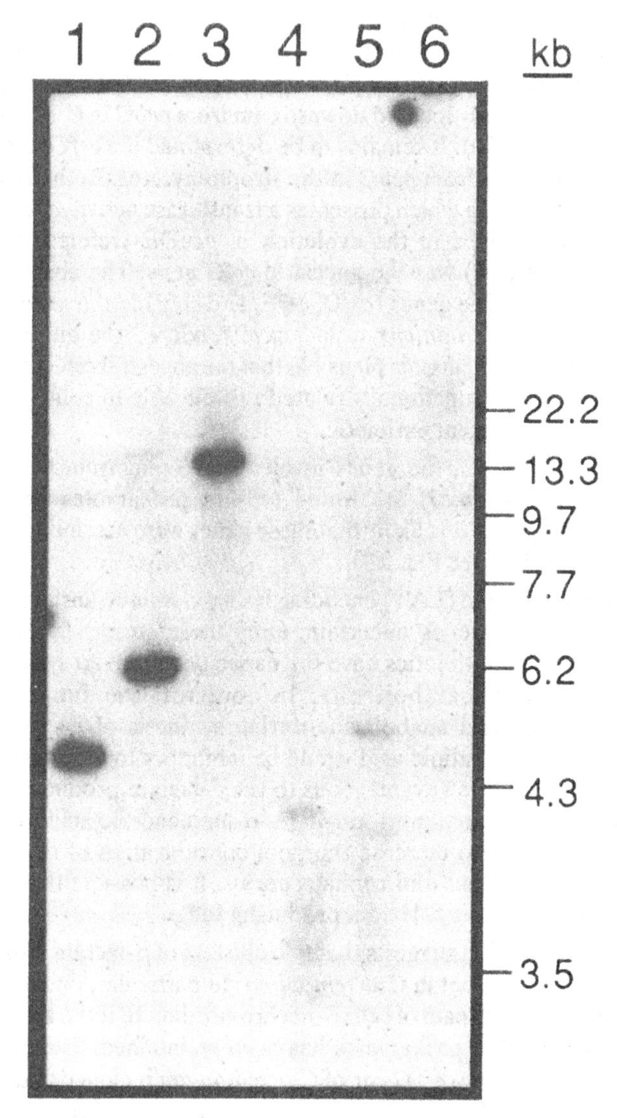

Fig. 9.4. Southern hybridization analysis confirmed the presence of related DNA sequences upstream of *pcbC* in *C. acremonium* and *P. chrysogenum*. Genomic DNA samples were cut with restriction endonucleases, DNA fragments were separated by agarose gel electrophoresis, transferred to Zeta-Probe membrane (BioRad) and hybridized with a probe radiolabelled by the polymerase chain reaction (Schowalter & Summer, 1989). The probe was a 3·1 kb fragment of DNA including the region from approximately 3·4 to 6·5 kb upstream of *pcb*C in *C. acremonium*. Lanes 1-3 contain *C. acremonium* DNA and Lanes 4-6 contain *P. chrysogenum* DNA. The DNA in lanes 1 & 4 was cut with *Bam*HI, DNA in lanes 2 & 5 was cut with *Bgl*II, and DNA in lanes 3 & 6 was cut with *Eco*RI.

hypothesis would suggest that if a *cef*G-like gene were linked to other β-lactam biosynthetic genes in the ancestral cephalosporin producer, then it should be found in a similar physical position in *C. acremonium*. The *cef*G gene was recently located downstream from *pcb*C in *C. acremonium* (Ramsden *et al*., 1990). It remains to be determined if a *cef*G-like gene is located downstream from *pcb*C in the streptomycetes. Both *pen*DE and *cef*G encode a protein which possesses a transferase activity. Perhaps the vestigial gene involved in the evolution of *pen*DE (referred to in the preceding paragraph) was the ancestral *cef*G gene. The absence of the cephalosporin specific genes (*cef*D, *cef*E, and *cef*F) in the ancestor of *P. chrysogenum* and *A. nidulans* would have rendered the ancestral *cef*G gene nonfunctional. Thus it is plausible that the ancestral *cef*G gene would evolve to produce a functionally related protein able to convert isopenicillin N to a more potent antibiotic.

In contrast to *cef*G, the genes involved in cephamycin C formation (*cmc*H, *cmc*I, and *cmc*J) are found only in prokaryotes such as the streptomycetes. Thus it is likely that these genes were assembled after the horizontal transfer (see Fig. 9.3).

The fate of the gene (LAT) encoding lysine-ε-aminotransferase during the horizontal transfer is uncertain. Only those streptomycetes which produce β-lactam antibiotics have the capacity to convert lysine to L-α-aminoadipic acid catabolically. In contrast, the fungi produce L-α-aminoadipic acid anabolically during synthesis of lysine. Limited supply of L-α-aminoadipic acid would be inhibitory to β-lactam production. Thus it would be advantageous to the β-lactam producing fungi to utilize a catabolic reaction producing L-α-aminoadipic acid during secondary metabolism to increase the pool concentration of this essential precursor of penicillins and cephalosporins. It is possible that the LAT gene does reside in the β-lactam producing fungi.

Another prediction suggests that two clusters of β-lactam biosynthetic genes should be present in *C. acremonium*. In particular, one may expect to find *cef*D just upstream of *cef*EF on chromosome II, if the arrangement of genes observed in prokaryotes has been maintained. This prediction will be tested when the *cef*D gene of *C. acremonium* is cloned and mapped.

The other cluster of β-lactam biosynthetic genes will reside on chromosome VI in *C. acremonium*. The ACV synthetase gene (*pcb*AB) of *C. acremonium* should be located in a position analogous to *pcb*AB in *A. nidulans* and *P. chrysogenum*. Therefore, it should be located upstream of *pcb*C on chromosome VI of *C. acremonium*. This prediction was tested in this study by the following experiments.

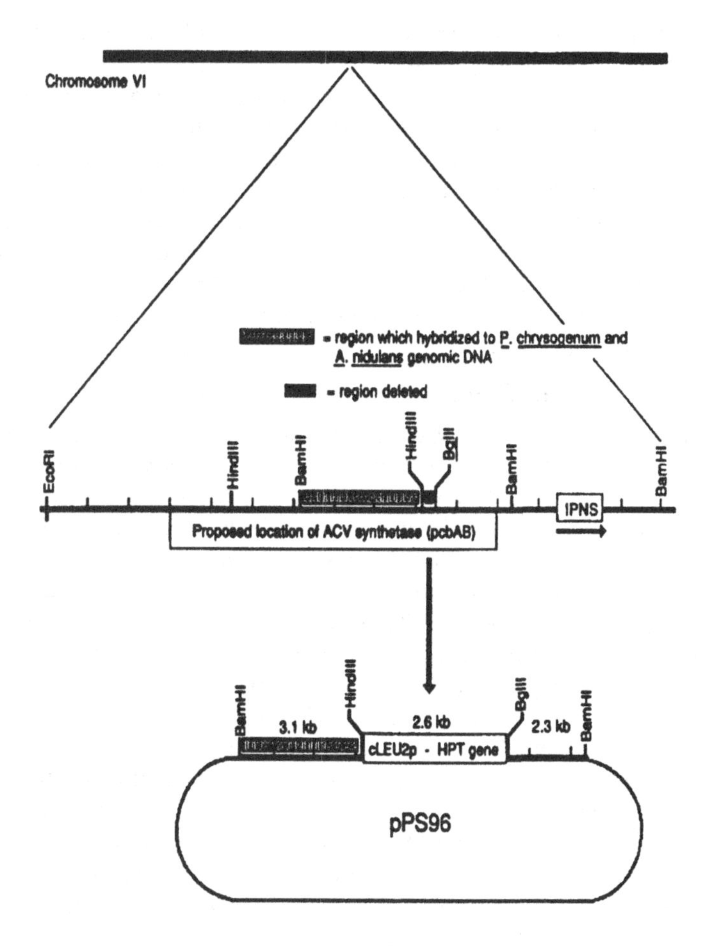

Fig. 9.5. Construction of plasmid pPS96.

Cross-species hybridization studies

If *pcb*AB resides upstream of *pcb*C in the fungi then one would expect *P. chrysogenum* and *C. acremonium* to share about 73% nucleotide identity (Fig. 9.2) at some point in this region. To test this, a 3·1 kb *Bam*HI to *Hin*dIII restriction fragment located upstream of *pcb*C in *C. acremonium* (Fig. 9.5) was radiolabelled, and used as a probe in a Southern analysis of genomic DNA from *P. chrysogenum* and *C. acremonium* (Fig. 9.4). As anticipated, strong hybridization to the appropriate restriction fragments in DNA from *C. acremonium* was observed. Less intense bands of hybridization occurred in lanes containing *P. chrysogenum* DNA. The size of the hybridizing restriction fragments in *P. chrysogenum* DNA were consistent with the size of restriction fragments upstream of *pcb*C in *P. chrysogenum*. When *A. nidulans* chromosomal DNA was subjected to the same test, still less intense hybridization was observed (data not shown). This pattern of diminished response might be anticipated based on DNA identities observed between the *pcb*C genes cloned from these organisms (Fig. 9.2). These data suggest the presence of a gene which is structurally related in these organisms upstream of *pcb*C. However, the identity of the common gene is unknown. Functional identity of this genomic region in *C. acremonium* was established by the following gene disruption experiment.

Targeted disruption of the genomic region upstream of *pcb*C in *C. acremonium* by transformation

A plasmid vector (pPS96) was designed and constructed to disrupt the genomic region upstream of *pcb*C in *C. acremonium* by homologous integration (Fig. 9.5). The ACV synthetase protein has been purified from *A. nidulans*. The size of this protein was estimated to be 220 kDa (van Liempt, von Dohren & Kleinkauf, 1989). Estimates of the size of ACV synthetase from *C. acremonium* are slightly larger (O'Callaghan, unpublished results). Thus, the *pcb*AB gene may be at least 10·0 kb in length. A 5·4 kb *Bam*HI restriction fragment, obtained from the region upstream of *pcb*C, was cloned into pUC19. A small internal *Hin*dIII-*Bgl*II restriction fragment was deleted and replaced with a hygromycin B phosphotransferase (HPT) gene fused in frame with the LEU2 promoter of *C. acremonium*. The HPT gene provided a dominant selection for transformation of *C. acremonium*. A bioassay, designed to detect production of β-lactam antibiotics, was performed on transformants of *C. acremonium* obtained with plasmid pPS96. Approximately 3% of the transformants recovered were unable to produce a β-lactam antibiotic. Selected transformants were studied in more detail.

Southern analysis

Homologous integration of plasmid pPS96 upstream of *pcb*C might occur either by a single or double cross-over event. Disruption of *pcb*AB function could occur in either event because the entire *pcb*AB gene was not represented on plasmid pPS96. In the case of a double cross-over, the endogenous 5·4 kb *Bam*HI fragment would disappear and a hybridizing fragment would appear at greater than 7·0 kb. Hybridization analysis of genomic DNA containing a single homologous cross-over should reveal the presence of the 5·4 kb *Bam*HI fragment plus other hybridizing bands corresponding to the integrated plasmid. To test these possibilities, total DNA was isolated from the untransformed recipient strain and transformants. Hybridization analysis with a probe specific for the 5·4 kb DNA fragment revealed the presence of this restriction fragment in DNA from

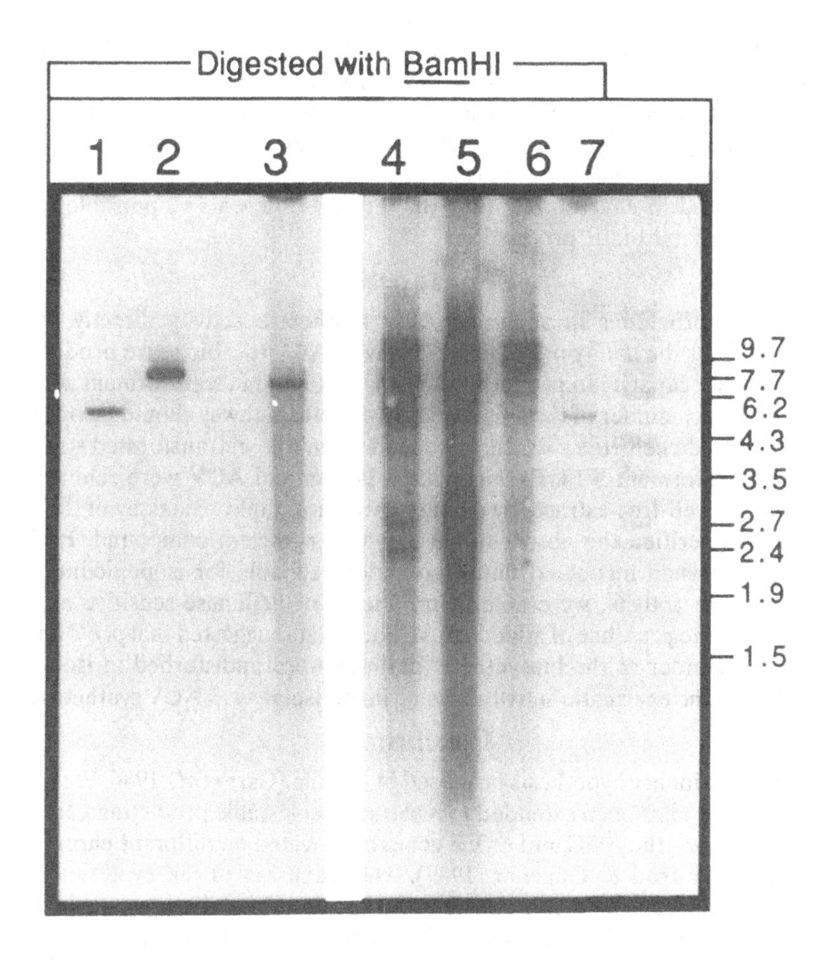

Fig. 9.6. Southern analysis of total DNA from transformants obtained with plasmid pPS96. DNA fragments obtained by digestion with the restriction enzyme *Bam*HI were separated by agarose gel electrophoresis, transferred to Zeta-Probe membrane and hybridized with probe radiolabelled by the polymerase chain reaction (Schowalter & Sommer, 1989). Lane 1: a purified sample of the 5·4 kb *Bam*HI fragment which lies upstream of *pcb*C in *C. acremonium*. Lane 2: pPS96 plasmid DNA. Lane 3: DNA from transformant #1 (β-lactam$^-$). Lane 4: DNA from transformant #16 (β-lactam$^-$). Lane 5: DNA from transformant #18 (β-lactam$^-$). Lane 6: DNA from transformant #20 (β-lactam$^+$). Lane 7: DNA from the untransformed recipient strain (β-lactam$^+$) used in the transformation experiment. Lanes 1-3 were exposed to film for 30 min. Lanes 4-7 were exposed to film for 3 hr.

the untransformed strain (Fig. 9.6, lane 7). All four transformants analyzed (Fig. 9.6, lanes 3-6) revealed the presence of multiple integration events (perhaps with tandem reiteration). In addition, a band of hybridization was evident in all transformants comigrating with the 5·4 kb *Bam*HI fragment in the untransformed strain. These results did not support a gene replacement event. Rather they were consistent with a single cross-over integration event. Further analysis of transformants will be required to characterize fully the integration events responsible for removal of antibiotic production.

Enzyme activity

Due to difficulties in analyzing ACV synthetase activity directly, we assayed for the ability or inability to convert ACV to a bioactive product. If only the capacity to produce ACV was disrupted in transformant #18, then the remainder of the β-lactam biosynthetic pathway should be operative. Crude cell-free extracts were made from the untransformed strain and transformant #18. Endogenous β-lactam and ACV were removed from the cell-free extracts by column chromatography. Bioassay of these extracts verified the absence of a bioactive β-lactam compound. Both extracts, when incubated under conditions suitable for isopenicillin N synthetase activity, were able to produce a penicillinase-sensitive antibiotic in the presence of added ACV. These data suggested that *pcb*C and the remainder of the biosynthetic pathway were undisturbed in isolate #18 and the enzymatic activity missing in this isolate was ACV synthetase.

Conclusions

An evolutionary hypothesis proposed by Ingolia (Carr *et al.*, 1986; Weigel *et al.*, 1988) has been extended to make several testable predictions. In *C. acremonium* the *pcb*C and *cef*EF genes are located on different chromosomes (Skatrud & Queener, 1989). The extension of the evolutionary hypothesis suggested two clusters of β-lactam biosynthetic genes in *C. acremonium*. Evidence presented for the presence of *pcb*AB upstream of *pcb*C on chromosome VI of *C. acremonium* lends support to the notion that two clusters of β-lactam genes exist in this fungus.

References

Cantwell, C. A., Beckman, R. J., Dotzlaf, J. E., Fisher, D. L., Skatrud, P. L., Yeh, W. -K. & Queener, S. W. (1990). Cloning and expression of a hybrid *Streptomyces clavuligerus cef*E gene in *Penicillium chrysogenum*. *Current Genetics* **17**, 213-221.

Carr, L. G., Skatrud, P. L., Scheetz, M. E., Queener, S. W. & Ingolia, T. D. (1986). Cloning and expression of the isopenicillin N synthetase gene from *Penicillium chrysogenum*. *Gene* **48**, 257-266.

Chen, C. W., Lin, H. -F., Kuo, C. L., Tsai, H. -L. & Tsai, F. -Y. (1988). Cloning and expression of a DNA sequence conferring cephamycin C production. *Bio/Technology* **6**, 1222-1224.

Jensen, S. E. (1985). Biosynthesis of cephalosporins. *Critical Reviews of Biotechnology* **3**, 277-301.

Ingolia, T. D. & Queener, S. W. (1989). Beta-Lactam biosynthetic genes. *Medicinal Research Reviews* **9**, 245-264.

Kovacevic, S., Weigel, B. J., Tobin, M. B., Ingolia, T. D. & Miller, J. R. (1989). Cloning, characterization, and expression in *Escherichia coli* of the *Streptomyces clavuligerus* gene encoding deacetoxycephalosporin C synthetase. *Journal of Bacteriology* **171**, 754-760.

Kovacevic, S., Tobin, M. B. & Miller, J. R. (1990). The β-lactam biosynthesis genes for isopenicillin N epimerase and deacetoxycephalosporin C synthetase are expressed from a single transcript in *Streptomyces clavuligerus*. *Journal of Bacteriology* **7**, 3952-3958.

Madduri, K., Stuttard, C. & Vining, L. C. (1990). Cloning of a gene governing lysine-ε-aminotransferase, the first enzyme in beta-lactam biosynthesis in *Streptomyces*. *Journal of Cellular Biochemistry, supplement* **14A**, 108.

Miller, J. R. & Ingolia, T. D. (1989). Cloning β-lactam genes from *Streptomyces* spp. and fungi. In *Genetics and Molecular Biology of Industrial Microorganisms*, (ed. C. L. Hershberger, S. W. Queener & G. Hegeman), pp. 262-269. American Society for Microbiology: Washington, D.C.

Queener, S. W. & Neuss, N. (1982). The biosynthesis of β-lactam antibiotics. In *The Chemistry and Biology of β*-lactam Antibiotics, (ed. R. B. Morin & M. Morgan), pp. 1-81. Academic Press, Inc.: London & New York.

Queener, S. W., Ingolia, T. D., Skatrud, P. L., Chapman, J. L. & Kaster, K. R. (1985). A system for genetic transformation of *Cephalosporium acremonium*. In *Microbiology-1985*, (ed. L. Leive), pp. 468-472. American Society for Microbiology: Washington DC.

Ramon, D., Carramolino, L., Patino, C., Sanchez, F. & Penalva, M. A. (1989). Cloning and characterization of the isopenicillin N synthetase gene mediating the formation of the β-lactam ring in *A. nidulans*. *Gene* **57**, 171-181.

Ramsden, M., Bradley, C., Scrogham, A. & Spence, D. (1990). Identification and cloning of the *cef*G gene of *A. chrysogenum*. In *Abstracts, 6th International Symposium on Genetics of Industrial Microorganisms* (August 12-18, 1990), p. 132.

Samson, S. M., Belagaje, R., Blankenship, D. T., Chapman, J. L., Perry, D., Skatrud, P. L., VanFrank, R. M., Abraham, E. P., Baldwin, J. E., Queener, S. W. & Ingolia, T. D. (1985). Isolation, sequence determination and expression in *Escherichia coli* of the isopenicillin N synthetase gene from *Cephalosporium acremonium*. *Nature* **318**, 191-194.

Samson, S. M., Dotzlaf, J. E., Slisz, M. L., Becker, G. W., Van Frank, R. M., Veal, L. E., Yeh, W. -K., Miller, J. R. Queener, S. W. & Ingolia, T. D. (1987). Cloning and expression of the fungal expandase/hydroxylase gene involved in cephalosporin biosynthesis. *Bio/Technology* **5**, 1207-1214.

Schowalter, D. B. & Sommer, S. S. (1989). The generation of radiolabeled DNA & RNA probes with polymerase chain reaction. *Analytical Biochemistry* **177**, 90-94.

Seno, E. T. & Baltz, R. H. (1989). Structural organization and regulation of antibiotic biosynthesis and resistance genes in *Actinomycetes*. In *Regulation of Secondary Metabolism in Actinomycetes*, (ed. S. Shapiro), pp. 1-73 CRC Press Inc.: Boca Raton, Florida.

Shiffman, D., Mevarech, M., Jensen, S. E., Cohen, G. & Aharonowitz, Y. (1988). Cloning and comparative sequence analysis of the gene coding for isopenicillin N synthetase in *Streptomyces*. *Molecular and General Genetics* **214**, 562-569.

Skatrud, P. L., Fisher, D. L., Ingolia, T. D. & Queener, S. W. (1986). Improved transformation of *Cephalosporium acremonium*. In *Genetics of Industrial Microorganisms*, (ed. M. Alacevic, D. Hranueli & D. Toman), pp. 111-119. American Society of Microbiology: Washington, DC.

Skatrud, P. L., Ingolia, T. D., Fisher, D. J., Chapman, J. L., Tietz, A. & Queener, S. W. (1989). Use of recombinant DNA technology to improve cephalosporin C production in *C. acremonium*. *Bio/Technology* **7**, 477-485.

Skatrud, P. L. & Queener, S. W. (1989). An electrophoretic molecular karyotype for an industrial strain of *Cephalosporium acremonium*. *Gene* **78**, 331-338.

Skatrud, P. L., Queener, S. W., Carr, L. G. & Fisher, D. L. (1987). Efficient integrative transformation of *Cephalosporium acremonium*. *Current Genetics* **12**, 337-348.

Smith, D. J., Burnham, M. K. R., Edwards, J., Earl, A. J. & Turner, G. (1990). Cloning and heterologous expression of the penicillin biosynthetic gene cluster from *Penicillium chrysogenum*. *Bio/Technology* **8**, 39-41.

Tobin, M. B., Fleming, M. D., Skatrud, P. L. & Miller, J. R. (1990). The acylCoA:Isopenicillin N acyltransferase genes (*pen*DE) from *Penicillium chrysogenum* and *Aspergillus nidulans*: molecular characterization and activity of recombinant enzyme in *Escherichia coli*. *Journal of Bacteriology* **172**, in press.

Van Liempt, H., von Dohren, H. & Kleinkauf, H. (1989). δ-(L-α-aminoadipyl)-L-cysteinyl-D-valine synthetase from *Aspergillus nidulans*. *Journal of Biological Chemistry* **264**, 3680-3684.

Veenstra, A. E., Van Solingen, P., Huininga-Muurling, H., Koekman, B. P., Groenen, M. A. M., Smaal, E. B., Kattevilder, A., Alvarez, E., Barredo, J. L. & Martin, J. F. (1989). Cloning of penicillin biosynthetic genes. In *Genetics and Molecular Biology of Industrial Microorganisms*, (ed. C. L. Hershberger, S. W. Queener & G. Hegeman), pp. 262-269. American Society for Microbiology: Washington, D.C.

Weigel, B. J., Burgett, S. G., Chen, V. J., Skatrud, P. L., Frolik, C. A., Queener, S. W. & Ingolia, T. D. (1988). Cloning and expression in *Escherichia coli* of isopenicillin N synthetase genes from *Streptomyces lipmanii* and *Aspergillus nidulans*. *Journal of Bacteriology* **170**, 3817-3826.

Chapter 10

Applications of genetically manipulated yeasts

Alexander W. M. Strasser, Zbigniew A. Janowicz, Rainer O. Roggenkamp, Ulrike Dahlems, Ulrike Weydemann, Armin Merckelbach, Gerd Gellissen, R. Jürgen Dohmen, Michael Piontek, Karl Melber & Cornelis P. Hollenberg

Tissue culture cells, bacteria and yeast are currently used as expression systems for the production of proteins valuable to industry. Yeast systems (based mainly on *Saccharomyces cerevisiae*) are characterized by good productivity, a lack of endotoxins and an established efficient fermentation technology. In addition, *S. cerevisiae* is genetically well characterized, and is thus most often used as a host organism.

In contrast to *Escherichia coli*, yeast possesses post-translational processing systems similar to those of mammalian cells. However, the system needs to be improved so as to provide higher cell productivity, increase secretion potential and enable defined post-translational modifications in protein processing and glycosylation to be carried out. To fulfil these requirements we have developed a system using the methylotrophic yeast species *Hansenula polymorpha*. Growth of this strain on methanol as sole energy and carbon source, is accompanied by strong induction of the enzymes involved in methanol metabolism (e.g. methanol oxidase, formate dehydrogenase or dihydroxyacetone synthase) and by the appearance of special organelles called peroxisomes. This induction mechanism shows features favourable for biotechnological application.

In addition to the production of foreign proteins, genetically manipulated yeasts can be used to improve established biotechnological processes, such as brewing, baking and ethanol production. In all these processes the compounds of interest are produced using disaccharides or glucose as a carbon source for metabolism. The organism most widely used in these processes is *S. cerevisiae* which is able to convert only sugar compounds into ethanol. Polysaccharides, starch, cellulose or hemicellulose, which constitute the majority of carbohydrate plant biomass cannot be utilized as fermentation substrates by *S. cerevisiae*. At the moment, metabolism of such compounds depends on addition of enzymes (e.g. α-amylase, glucoamylase, xylose isomerase, endoglucanase, exoglucanase and β-glucanase) prior to fermentation. Thus, it is of great commercial interest to develop new yeast strains which can convert such polysaccharides directly into fermentable sugars.

In brewing, the fermentable sugars are provided from barley by partial hydrolysis of starch during the malting process. This process, however, results in production of considerable amounts or unfermentable carbohydrates like dextrins. These contribute to the high calorie content of beer manufactured by traditional methods and thus have to be removed in the production of diet beer. The present production process depends on addition of glucoamylase and the product most widely used currently is derived from *Aspergillus* species. However, this enzyme cannot be inactivated at temperatures normally employed in breweries for pasteurization. This can result in an unstable beer which changes its chemical properties during storage.

A newly developed yeast strain with amylolytic properties provides a tool for the production of beer, especially light beer, which does not depend on the availability of expensive industrial enzymes or even malt. The application of an amylolytic baker's yeast in baking could also eliminate the need for α-amylase-enriched flour in the manufacture of certain types of bread.

Hansenula polymorpha as an expression system Schwanniomyces *glucoamylase secretion*

When a transformation system for *H. polymorpha* was established (Roggenkamp *et al.*, 1986) this methylotrophic yeast became available for gene technology. We have investigated its methanol metabolism, and have cloned and characterized the genes involved (Ledeboer *et al.*, 1985; Janowicz *et al.*, 1985). The strong promoters of these genes (methanol oxidase and formate dehydrogenase) can be used for the efficient production of foreign proteins in *H. polymorpha*. The glucoamylase gene from *Schwanniomyces occidentalis* containing its authentic signal sequence was inserted into an *H. polymorpha* expression/integration vector under the control of the methanol-inducible formate dehydrogenase promoter. Transformation of a competent *H. polymorpha* strain resulted in mitotically stable transformants which secrete active glucoamylase into the medium. Analysis of the secreted enzyme showed that the heterologous protein was faithfully synthesized, processed and modified. The N-terminus of the secreted polypeptide and the extent of glycosylation are identical in *S. occidentalis* and in *H. polymorpha*. In fermentations with the transformed strains under inducing conditions a yield of $1{\cdot}4$ g l^{-1} of secreted enzyme was obtained at a culture density of 130 g dry weight per litre.

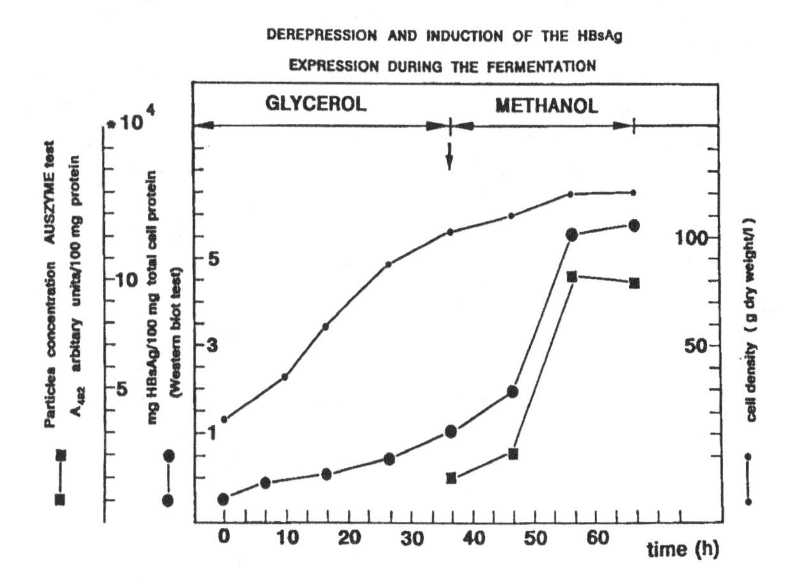

Fig.10.1. Growth of HBsAg expressing transformants of *Hansenula polymorpha* on non selective synthetic medium. At the timepoint indicated by the vertical arrow glycerol supply (derepression of the promoter) was cut off. The subsequent addition of methanol causes induction of expression. The amount of antigenic protein was measured by Western blots (circles). The amount of 22 nm particles produced was estimated by the AUSZYME (Abbott) test (squares).

Thus, *H. polymorpha* can be a very efficient, mitotically stable organism for production of heterologous secretory proteins.

Production of particles containing Hepatitis B virus antigens

Our studies with the hepatitis B surface antigen (HBsAg), which is expressed under the control of the formate dehydrogenase promoter, have confirmed that the physiological and genetic features of *H. polymorpha* make this organism a very efficient, mitotically stable production system, which is stringently controlled.

About 5% of the total cell protein can be identified as HBsAg (Fig. 10.1). Moreover, 90% of synthesized protein is present in the form of highly antigenic 22 nm particles. Particles containing S-protein synthesised in *Saccharomyces cerevisiae* have been successfully used by other authors for the formulation of a recombinant vaccine (Valenzuela *et al.*, 1982; Harford *et al.*, 1983). As shown in Fig. 10.1, a special feature of *H. polymorpha* is its ability to grow vigorously on glycerol. The very low production of ethanol from this carbon source is advantageous for the fermentation parameters. HBsAg is produced with about 20 to 30% of its maximal efficiency under these conditions because of the derepression of the formate dehydrogenase promoter. The exhaustion or the first carbon source followed by addition of methanol causes a rapid induction of the expression. Thus, together with the high biomass production, this expression system facilitates the economical isolation of the desired foreign polypeptides which can be used as a vaccine.

In another set of experiments we have simultaneously expressed L (preS1-S2-S) and S Hepatitis B antigens in this system under the control of a strong methanol inducible promoter derived from the methanol oxidase gene. Several multimeric integrants were constructed containing various numbers of L- and S- expression cassettes. Thus, we have obtained a wide spectrum of strains characterized by different preS to S ratios. The expression level of the foreign proteins was 5 to 7% of the total cell protein.

Analysis by sucrose or CsCl gradient centrifugation and tests with particle-specific immunological reagents demonstrated that 90% of the expressed protein spontaneously formed sub-viral particles containing both proteins. Biochemical analysis revealed that this particle contained yeast-derived membrane lipids. This, together with the fact that 1 to 5% of the L protein is N-linked glycosylated (demonstrated by endoglycosidase H digestion studies) indicates that also in *H. polymorpha* these antigens have transmembrane properties. In animal cells HBsAg is synthesized as transmembrane proteins which possess two complex signal sequences, one near the N-terminus and one internal whose conjoint action determines the final transmembrane location in the membrane of the endoplasmic reticulum (Eble *et al.*, 1986).

Thus, we have developed a system, which produces composite particles bearing both S- and PreS1-specific epitopes. Such particles can be used for the production of a vaccine which might provide a wider spectrum of protection and possibly increase immunogenicity.

Overproduction of methanol oxidase

Methanol oxidase can be used as a bleaching agent in the production of detergents, for the analytical determination of alcohols, and for the elimination of traces of oxygen in food. Oxidase preparations are often contaminated with other enzymes; catalase contamination, in particular, severely decreases the functional value of such preparations. Therefore commercially available enzyme cannot be produced at low cost. Recently,

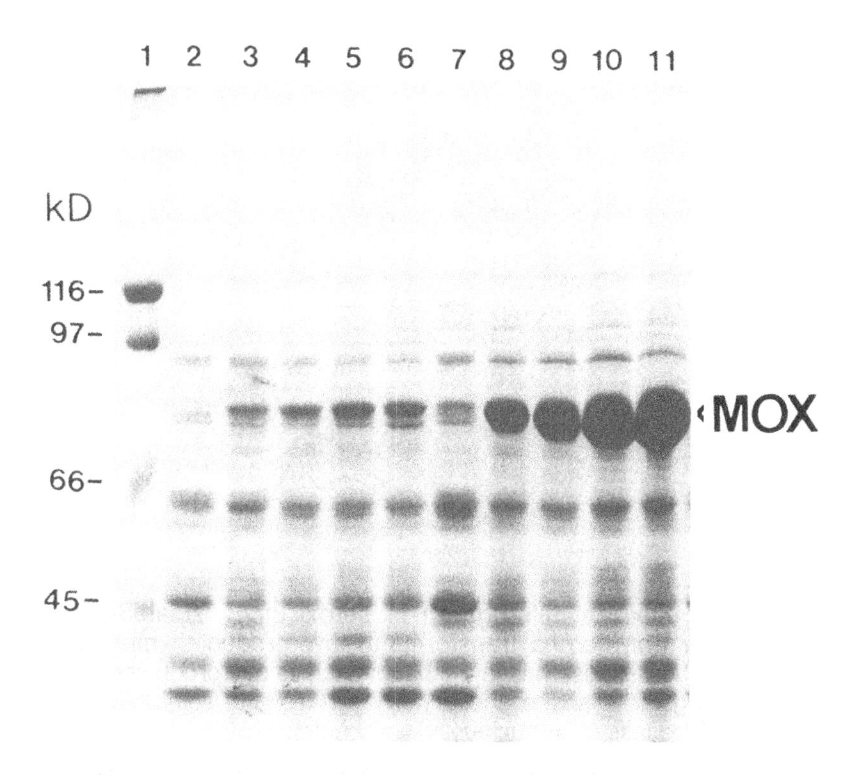

Fig. 10.2. Polyacrylamide gel electrophoresis of crude extracts from *Hansenula polymorpha* mutants transformed with the methanol oxidase gene (*MOX*) on a high copy number plasmid. Lane 1, size markers (kilodaltons); lanes 2 to 6 show extracts prepared from the recipient catalase mutant cells grown for 0·8, 16, 27, and 48 h. Lanes 7-11 show MOX production in the transformed mutant strain grown for the same periods.

we have isolated a mutant of *H. polymorpha* which is deficient in catalase and methanol oxidase expression (Roggenkamp, Didion & Kowalik, 1989). In addition, this mutant is deficient in the expression of orotidine 5' phosphate decarboxylase (*URA3*). Transformation of this mutant, using a plasmid containing the functional methanol oxidase gene, a sequence for multiple autonomous replication and *URA3* as a selective marker results in transformants which express active methanol oxidase in concentrations of up to 75% of total cellular protein (Fig. 10.2). The peroxisomes of these methanol oxidase overproducing cells (Fig. 10.3) show a drastic

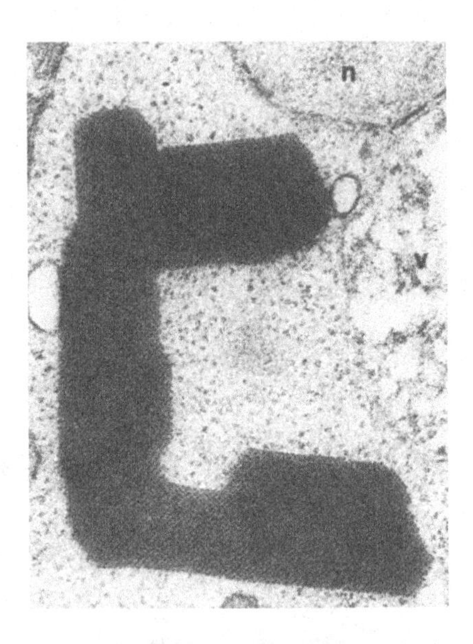

Fig. 10.3. Electron micrograph of peroxisomes isolated from *Hansenula polymorpha* transformants overproducing methanol oxidase. The peroxisomes in methanol oxidase overproducing cells are rectangular in shape and densely packed with semicrystalline methanol oxidase (n = nucleus; v = vacuole).

increase in size, are rectangular instead of round or oval in shape; and are densely packed with semicrystalline methanol oxidase. Initial results of studies using this strain for high level expression or genes other than methanol oxidase are promising.

Construction of amylolytic *Saccharomyces cerevisiae* strains

Cloning Schwanniomyces *genes and their expression in* S. cerevisiae

A prerequisite for the construction of amylolytic *Saccharomyces cerevisiae* strains by gene technology, is the availability of suitable genes coding for α-amylase and glucoamylase. After extensive screening of different amylolytic microorganisms we focused our interest on the yeast *Schwanniomyces occidentalis* (Price, Fuson & Phaff, 1978) which is able to grow on starch as a sole carbon source. *S. occidentalis* hydrolyses starch completely into glucose using two extracellular enzymes, α-amylase and glucoamylase (syn. amyloglucosidase), an enzyme with significant debranching activity (Dohmen *et al.*, 1989; Oteng Gyang, Moulin & Galzy,

1981; Sills & Steward, 1982; Sills, Sauder & Steward, 1984; Tubb, 1986). This amylolytic system is irreversibly inactivated at 60°C and is thus especially advantageous for applications in brewing and baking.

The α-amylase gene (*AMY1*), of *Schwanniomyces occidentalis* was isolated from a cosmid gene library by screening in *Saccharomyces cerevisiae* for α-amylase secretion. The *AMY1* gene was expressed from its original promoter in *Saccharomyces cerevisiae*, *Schizosaccharomyces pombe* and *Kluyveromyces lactis* (Strasser *et al.*, 1989), leading to an active secreted gene product.

Expression of this gene in *Saccharomyces cerevisiae* allows growth on starch as a sole carbon source but the amount of α-amylase protein in the cell free supernatant is relatively low. High expression of the *AMY1* gene in *Saccharomyces cerevisiae* was achieved after fusion of the *S. cerevisiae GAL10* promoter to the *AMY1* structural gene.

The *Schwanniomyces occidentalis* glucoamylase gene (*GAM1*) was isolated from a Charon 4A genomic library using synthetic oligonucleotides encoding tryptic peptides of the glucoamylase protein as a probe (Dohmen *et al.*, unpublished). In contrast to the *AMY1* gene, the *GAM1* gene is not expressed from its original promoter sequences in *Saccharomyces cerevisiae*. For expression studies the *S. cerevisiae GAL1* promoter was fused to the *GAM1* structural gene. This fusion led to expression and secretion of active glucoamylase in *S. cerevisiae*. The resultant transformants have acquired the ability to hydrolyse starch completely into glucose and are consequently able to grow on starch as a sole carbon source. Compared to transformants harbouring the *AMY1* gene, these transformants exhibit significantly better growth on starch.

Genetically engineered amylolytic S. cerevisiae

An expression cassette, containing both the *AMY1* and *GAM1* genes under control of the *GAL10* and *GAL1* promoters, respectively, was cloned into a centromere plasmid. Transformants obtained with this plasmid had one copy of each gene per haploid genome. A laboratory strain of *Saccharomyces cerevisiae* was transformed with this construct to study secretion (promoted by their authentic signal sequences) of both enzymes and to characterize the efficiency of the amylolytic system of *Schwanniomyces occidentalis* in *Saccharomyces cerevisiae*. The expression of both genes was found to be inducible by galactose. The activity of the secreted enzymes was analyzed in the cell-free supernatant and compared with the extracellular activity of the donor organism *Schwanniomyces occidentalis*. This comparison (Table 10.1) demonstrates that the amylolytic system is as efficient in the genetically engineered *Saccharomyces cerevisiae* as in the original *Schwanniomyces occidentalis*.

Table 10.1. Comparative measurements of the extracellular amylolytic activities of wild type *Schwanniomyces occidentalis* and genetically engineered *Saccharomyces cerevisiae*

Extracellular enzymes ($U\ l^{-1}$)	*Saccharomyces cerevisiae* genetically engineered	*Schwanniomyces occidentalis* wild type
Glucoamylase[a]	360	342
α-Amylase[b]	35	20

[a]one unit is the amount of enzyme that liberates 1 μ mol glucose from 5% soluble starch per min at 50°.

[b]one unit is the amount of enzyme that liberates 1 μ mol 2-chloro-4-nitrophenol per min at 37° using the Merckotest kit No. 14358

A stable integration of an analogous expression cassette into an industrial *Saccharomyces cerevisiae* yeast results in a strain which combines high fermentation rates and good ethanol tolerance with the newly acquired ability to degrade starch entirely (Strasser, 1988).

Conclusions

We have demonstrated the impact of genetically manipulated yeasts on food and pharmaceutical industries. The few examples show that this technology can influence significantly the alteration of nutritional and functional properties. This includes the development of food additives such as enzymes, the improvement of industrial strains and the establishment of economically competitive production systems for pharmaceutical compounds. Future advances in molecular biology and bioengineering will improve and extend the great potential for the application of genetically manipulated yeasts.

References

Dohmen, R. J., Strasser, A. W. M., Zitomer, R. S. & Hollenberg, C. P. (1989). Regulated overproduction of α-amylase by transformation of the amylolytic yeast *Schwanniomyces occidentalis*. *Current Genetics* **15**, 319-325.

Dohmen, R. J., Strasser, A. W. M., Dahlems, U. & Hollenberg, C. P. (1990). Cloning of the *Schwanniomyces occidentalis* glucoamylase-encoding gene (*GAM1*) and its expression in *Saccharomyces cerevisiae*. *Gene*, in press.

Eble, B., Lingappa, V. & Ganam, D. (1986). Hepatitis B surface antigen: an unusual secreted protein initially synthesized as a transmembrane polypeptide. *Molecular Cellular Biology* **6**, 1454-1463.

Harford, N., Cabezon, T., Crabeel, M., Simoen, E., Rutgers, T. & De Wilde, M. (1983). Expression of Hepatitis B surface antigen in yeast. *Developments in Biological Standardization* **54**, 125-130.

Janowicz, Z. A., Eckart, M. R., Drewke, C., Roggenkamp, R. O., Hollenberg, C. P., Maat, J., Ledeboer, A. M., Visser, C. & Verrips, C. T. (1985). Cloning and characterization of the DAS gene encoding the major methanol assimilatory enzyme from the methylotrophic yeast *Hansenula polymorpha. Nucleic Acids Research* **13**, 3043-3062.

Ledeboer, A. M., Edens, L., Maat, J., Visser, C., Bos, J. W., Verrips, C. T., Janowicz, Z. A., Eckart, M., Roggenkamp, R. O. & Hollenberg, C. P. (1985). Molecular cloning and characterization of a gene coding for methanol oxidase in *Hansenula polymorpha. Nucleic Acids Research* **13**, 3063-3082.

Oteng Gyang, K., Moulin, G. & Galzy, P. (1981). A study of the amylolytic system of *Schwanniomyces castellii. Zeitschrift für allgemeine Mikrobiologie* **21**, 537-544.

Price, C. W., Fuson, G. B. & Phaff, H. J. (1978). Genome comparison in yeast systematics: delimitation of species within the genera *Schwanniomyces, Saccharomyces, Debaryomyces* and *Pichia. Microbiological Reviews* **42**, 161-193.

Roggenkamp, R. O., Hansen, H., Eckart, M., Janowicz, Z. A. & Hollenberg, C. P. (1986). Transformation of the methylotrophic yeast *Hansenula polymorpha* by autonomous replication and integration vectors. *Molecular and General Genetics* **202**, 302-308.

Roggenkamp, R. O., Didion, T. & Kowallik, K. V. (1989). Formation of irregular giant peroxisomes by overproduction of the crystalloid core protein methanol oxidase in the methylotrophic yeast *Hansenula polymorpha. Molecular and Cellular Biology* **9**, 988-994.

Sills, A. M. & Steward, G. G. (1982). Production of amylolytic enzymes by several yeast species. *Journal of the Institute of Brewing* **88**, 313-316

Sills, A. M., Sauder, M. E. & Steward, G. G. (1984). Isolation and characterization of the amylolytic system of *Schwanniomyces castellii. Journal of the Institute of Brewing* **90**, 311-314.

Strasser, A. W. M., Selk, R., Dohmen, R. J., Niermann, T., Bielefeld, M., Seeboth, P., Tu, G. & Hollenberg, C. P. (1989). Analysis of the α-amylase gene of *Schwanniomyces occidentalis* and the secretion of its gene product in transformants of different yeast genera. *European Journal of Biochemistry* **184**, 699-706.

Strasser. A. W. M. (1988). Industrielle Nutzung von Mikroorganismen. *Bioengineering* **4**, 162-163.

Tubb, R. S. (1986). Amylolytic yeasts for commercial application. *Trends in Biotechnology* **4**, 98-103.

Valenzuela, P., Medina, A., Rutter, W. J., Ammerer, G. & Hall, B. D. (1982). Synthesis and assembly of Hepatitis B surface antigen particles in yeast. *Nature* **298**, 347-350.

Chapter 11

Molecular biology of fungal plant pathogenicity

Richard P. Oliver, Mark L. Farman, Nicholas J. Talbot & Mark T. McHale

Study of the molecular biology of fungal pathogenesis is at an exciting stage. In recent years, many technological hurdles have been passed and the first critical experiments linking specific genes to pathogenicity have been reported. In addition we have uncovered many surprises in the molecular biology of the organisms concerned. It is timely, therefore to review this rapidly moving field. In this review we will summarise the aims of the research, the techniques available and describe highlights of the achievements to date.

Plant pathogens are distinguished from saprotrophs by their ability to complete their life cycles and induce symptoms on living plants. This ability to cause disease may be termed basic pathogenicity. This definition raises several questions: do pathogens possess genes which saprotrophs lack or do pathogens lack genes which saprotrophs possess; is the regulation of such genes different in the two groups of fungi? The fact that many pathogens are specific for a particular plant species or genus indicates that there may be many different answers to these questions. Host plant specificity may be modified by a higher level of specificity whereby certain races of a pathogen are pathogenic only on particular cultivars of a host. This race-cultivar specificity is termed virulence. The concepts of pathogenicity and virulence are well illustrated by the pathogen *Cladosporium fulvum* (syn. *Fulvia fulva*) the cause of tomato leaf mould. *C. fulvum* is a non-obligate pathogen which infects a subset of the genus *Lycopersicon* (Table 11.1). Races of the pathogen exhibit a classical gene-for-gene pattern of interaction with cultivars of the cultivated tomato possessing different resistance genes (Table 11.2). Resistance genes were introgressed into tomato by hybridization with resistant species of *Lycopersicon*. Resistance of tomato to *C. fulvum* often involves recognition of the pathogen leading to a hypersensitive response. Failure to be recognised is a prerequisite to successful infection. Thus, it appears that races of the pathogen possess one or a few *avirulence* genes which prevent infection of cultivars containing the appropriate resistance genes. Quite why pathogens should possess these paradoxical avirulence genes is a major, long-standing question.

Table 11.1. Limited host range of *Cladosporium fulvum*

Susceptible hosts	Non-hosts
Lycopersicon esculentum var. Moneymaker	*L. peruvianum*
L. esculentum var. *cerasiforme*	*L. parviflorum*
Solanum (Lycopersicon) penelli	*L. hirsutum*
L. chmielski	*L. pimpinellifolium*

Data from Bond (1938), Kerr & Bailey (1964) and R. P. Oliver (unpublished).

The aims of molecular fungal plant pathology can be stated as the characterisation of genes controlling host plant specificity (pathogenicity), host-cultivar specificity (avirulence) and symptom production. This may appear too narrow in scope as our overall aim is to limit plant disease using environmentally acceptable techniques. With this in mind it becomes legitimate to focus on any fungal gene which contributes to the spread and success of the pathogen and not limit ourselves to the contrast with saprotrophs. Such studies might ultimately identify novel targets for fungicides.

Molecular biology of fungal pathogens

Transformation

An essential prerequisite for the analysis of fungal pathogenicity is the development of efficient vectors and protocols for transformation. Successful transformation has been reported for many species of pathogen and in only a few cases have significant efforts to establish the technique been unrewarded. A number of selectable markers have been used. (Table 11.3). Two strategies have found widespread success. The first is selection of nitrate assimilatory pathway mutants with chlorate, and complementation of the mutants with the *Aspergillus nidulans niaD* gene (Daboussi *et al.*, 1989; Malardier *et al.*, 1989). The second is the use of chimeric hygromycin phosphotransferase genes which link various fungal promoters to the *Escherichia coli* hph gene. One such plasmid, pAN7-1, with the *Aspergillus nidulans* GAPDH promoter (Punt *et al.*, 1987) has been used to transform more than twenty species of fungi including several pathogens (Table 11.4). This implies that the transcription initiation signals of this GAPDH promoter are sufficiently well recognised to be expressed in this range of fungi.

Transformation is achieved by PEG/Ca^{2+} treatment of protoplasts in almost all cases. Lithium acetate treatment of whole cells has been used

Table 11.2. Race-cultivar interactions of *Cladosporium fulvum* with tomato

Race	Tomato cultivar*							Fungal genotype[a]
	Cf0	Cf2	Cf4	Cf2.4	Cf5	CF9	CF11	
0	S[b]	R	R	R	R	R	R	A2, A4, A5, A9, A11
2	S	S	R	R	R	R	R	a2, A4, A5, A9, A11
4	S	R	S	R	R	R	R	A2, a4, A5, A9, A11
2,4	S	S	S	S	R	R	R	a2, a4, A5, A9, A11
5	S	R	R	R	S	R	R	A2, A4, a5, A9, A11
2,4,5	S	S	S	S	S	R	R	a2, a4, a5, A9, A11
2,4,9,11	S	S	S	S	R	S	S	a2, a4, A5, a9, a11
2,4,5,9,11	S	S	S	S	S	S	S	a2, a4, a5, a9, a11

*Indicates specific *C. fulvum* resistance genes (*Cf* genes) carried by cultivar.
[a]A2 indicates avirulence to *Cf* 2, a2 virulence to *Cf* 2, etc.
[b]S indicates susceptible, R indicates resistant.
Data from Bond (1938), Kerr & Bailey (1964), De Wit, Toma & Joosten (1988), P. J. G. M. De Wit (personal communication) and R. P. Oliver (unpublished).

to transform *Colletotrichum trifolii* (Dickman, 1988) and *Ustilago violacea* (Bej & Perlin, 1989), a procedure that is also suitable for *Leptosphaeria maculans* (M. L. Farman & R. P. Oliver, unpublished). Interestingly, this procedure appears to produce none of the so-called abortive transformants found with protoplast transformation. This observation suggests that formation of multinucleate cells during protoplast transformation may be required for the formation of the abortive colonies.

In almost all cases the DNA has been found to integrate in the genome. Transformation frequencies are correspondingly low, in the range 1 to 100 μg^{-1} DNA. The exception is *U. maydis*. Tsukuda *et al.* (1988) cloned random genomic fragments into an integration vector and found that 1 in 20 of the recombinants transformed at a frequency of 10^3 to 10^4 μg^{-1}. These vectors were shown to be autonomously replicating implying that, as in yeast, chromosomal replication origins are present in large numbers in the chromosomes of *U. maydis*.

Random integration of the transforming DNA has been reported in most cases. Homologous recombination and gene disruption have been described for *U. maydis* (Fotheringham & Holloman, 1989; Kronstad *et al.*, 1989). For most of the fungi under study, cloned homologous genes with which to investigate homologous recombination are not available. This problem can be circumvented by transforming a previously transformed strain, provided that both primary and secondary vectors show

Table 11.3. Selectable markers used for transformation of fungal plant pathogens

Marker	Species	Reference
neo^r	*Ustilago maydis*	Banks, 1983
amdS	*Cochliobolus heterostrophus*	Turgeon, Garber & Yoder, 1985
hyg^r	*Cladosporium fulvum*	Oliver *et al.*, 1987
hyg^r	*Cochliobolus heterostrophus*	Turgeon, Garber & Yoder, 1987
argB	*Magnaporthe grisea*	Parsons, Chumley & Valent, 1987
ben^r	*Colletotrichum trifolii*	Dickman, 1988
ben^r	*Gaeumannomyces graminis*	Henson, Blake & Pilgeram, 1988
pyr3	*Ustilago maydis*	Banks & Taylor, 1988
niaD	*Fusarium oxysporum*	Malardier *et al.*, 1989
pyr6	*Ustilago maydis*	Kronstad *et al.*, 1989
leu1	*Ustilago maydis*	Fotheringham & Holloman, 1989
phleo^r	*Cladosporium fulvum*	I. N. Roberts & R. P. Oliver, unpublished
phleo^r	*Leptosphaeria maculans*	M. L. Farman & R. P. Oliver unpublished

sequence homology. The phleomycin resistance vector pAN8-1 shares a pUC backbone and fungal transcription signals with pAN7-1 (Mattern, Punt & van den Hondel, 1988). Transformation of pAN7-1 into a single-copy pAN8-1 transformant of *L. maculans* results in a consistent doubling of transformation frequency. Furthermore, 4 out of 5 transformants had pAN7-1 integrated into pAN8-1 sequences (M. L. Farman & R. P. Oliver, unpublished). These results suggest that targeted integration is feasible even in genetically poorly-characterised species. One application would be in analysis of promoter deletions where targeted integration would eliminate chromosomal position effects.

Co-transformation of non-selected plasmids along with the selected marker has been reported frequently (e.g. Roberts *et al.*, 1989). Our studies with *L. maculans* indicate that co-integration of both vectors is a frequent feature of co-transformation and supports the view that the provision of sites of homology facilitates integration.

Use of reporter genes

Development of reporter genes will facilitate analysis of the transcriptional and translational control of gene expression. Promoters of genes of interest are ligated to the reporter gene and expression of the reporter gene, whose product is easily assayed, gives information about the activity of the promoter under different conditions. An early example of the use of promoter probe vectors led to the development of a chimeric hygromycin phosphotransferase gene as a selectable marker (Turgeon, Garber

Table 11.4. Species of pathogenic fungi transformed to hygromycin resistance with pAN7-1 (Punt *et al.*, 1987)

Species	Reference
Cladosporium fulvum	Oliver *et al.*, 1987
Leptosphaeria maculans	Farman & Oliver, 1988
Septoria nodorum	Cooley *et al.* 1988
Claviceps purpurea	Comino *et al.* 1989
Curvularia lunata	Osiewacz & Weber, 1989
Pseudocercosporella herpotrichoides	Blakemore *et al.* 1989
Trichosporon cutaneum	Glumoff *et al.* 1989
Ascochyta pisi	J. K. Bowen, pers. comm.
Mycosphaerella pinodes	J. K. Bowen, pers. comm.
Cryphonectria parasitica	Churchill *et al.* 1990

& Yoder, 1987). Random fragments of *Cochliobolus heterostrophus* DNA were cloned upstream of a promoterless hygromycin gene. The library was transformed into *C. heterostrophus* and the products screened for expression of hygromycin resistance. The chimeric gene was rescued from the few resistant colonies and used to develop an efficient vector.

The β-glucuronidase (GUS) gene is an ideal reporter gene for many systems but particularly plants and fungi (Jefferson, Kavanagh & Bevan 1987; Roberts *et al.*, 1989). Firstly, all fungi and plants studied to date lack endogenous GUS activity (unlike β-galactosidase) facilitating the detection of even low levels of GUS in transformed tissues. Secondly GUS is robust and readily assayed spectrophotometrically, fluorometrically or detected using histochemical stains.

An unexpected offshoot of these studies is the usefulness of transformed strains which constitutively express GUS and hence are easily detected by virtue of their GUS activity. This can be used in population biology studies to follow the spread of marked strains. It is also a convenient way to detect fungus within a plant, in fact it may be easier to assay putatively infected tissue for GUS than to observe infection microscopically. This facet has been refined to follow *L. maculans* as it progresses through the tissues of the host brassica. Sections of tissue were infiltrated with the histochemical stain X-gluc giving a rapid, non-destructive localisation of the fungus (M. L. Farman, unpublished).

Molecular karyotyping

Nearly all plant pathogenic fungi are poorly characterised genetically. The techniques of pulsed-field gel electrophoresis which separate chromosome-sized DNA molecules offer the opportunity to establish rapidly the karyotype of these organisms and to map cloned genes to particular chromosomes (Orbach *et al.*, 1988). This represents a major advance over the traditional, time-consuming strategy of linkage analysis. We have utilised this approach to produce a karyotype for *C. fulvum*. The chromosomal DNA resolved into nine bands estimated to range in size between 1·8 and 5·4 Mbp (The upper sizes in this range are speculatively based on the assumption of a linear relationship between mobility and size). The two largest bands are probably doublets giving a genome size of around 43 Mbp. The gel was blotted and hybridised successively to a series of probes to establish their chromosomal location. These techniques have great potential for genome mapping and to elucidate the nature of the variation between races of a pathogen. Pulsed-field gel electrophoresis will have an application in aiding the classification of fungal species, particularly in the case of imperfect pathogens, where the electrophoretic karyotype may well emerge as the most meaningful primary taxonomic criterion.

Development of new races of pathogens - a consequence of transposon activity

One of the most interesting and economically important features of fungal pathogens is the frequency and rapidity with which new resistant cultivars of a crop-plant become infected with (presumably) novel races of a pathogen. The ability to evolve virulent races gives rise to the boom-and-bust cycles of agriculture. The frequency with which novel specificities arise can be as a high 10^{-3} per generation (Dinoor, Eshed & Nof, 1988). The molecular basis of this production of novel races is unknown but this frequency is suggestive of the presence of transposable elements. Involvement of transposable elements in the evolution of new virulences in bacterial pathogens has been elegantly demonstrated (Kearney *et al.*, 1988). Until recently, direct evidence for transposable elements in any fungal pathogen was lacking. A procedure aimed at isolating fungal genes whose products are excreted into the apoplast has led rather serendipitiously to the discovery of a class of retrotransposon in *C. fulvum* (McHale *et al.*, 1989).

The genome of *C. fulvum* includes genomic copies of a moderately long terminal repeat retrotransposon which is similar to the *Drosophila* elements 17.6 and 297, a Lily retrotransposon *del* and the yeast Ty3 (Saigo *et al.*, 1984; Smyth *et al.*, 1989; Hansen, Chalker & Sandmeyer, 1988). A pair of identical long terminal repeats (LTRs) have been sequenced and shown to possess target-site duplications, inverted terminal duplications and + and – strand primer binding sites. Reverse transcriptase is detectable in homogenates of mycelium and copurifies with 40 nm diameter virus-like particles. The element (Cft-1) has been shown by restriction mapping and pulsed-field gels to be in different chromosomal locations (M. T. McHale & I. N. Roberts, unpublished).

The implications of these findings are many. The presence of transposable elements may explain the generation of novel races either via direct insertion and disruption of avirulence genes or by providing dispersed sites of homology around the genome promoting chromosomal translocations and deletions. The element may form the basis of sophisticated molecular genetic tools for the analysis of gene function not just in pathogens but possibly in all filamentous fungi. It seems very unlikely that *C. fulvum* will turn out to be the only filamentous fungus with a LTR-retrotransposon and preliminary evidence suggests that a wide range of other filamentous species have related elements. This raises the further question of the origin of these elements. The similarity of Cft-1 to elements in species as diverse as yeast, *Drosophila* and lily, yet distinct from other yeast and *Drosophila* retrotransposons is strongly suggestive of horizontal transmission of genetic information across species barriers. The alternative hypothesis of convergent molecular evolution seems even more unlikely at present. Further dissection of such ideas will require much detailed molecular evidence on the structure of these elements in ecologically and taxonomically related species.

Molecular analysis of pathogenicity genes

The approaches which have been adopted to clone pathogenicity genes can be broadly classified into two groups - targeted and non-targeted. In the former, specific genes whose products are believed *a priori* to be involved in pathogenicity are sought directly: in the latter, the view was broader, the aim being to clone any gene, predictable or not, involved in pathogenicity. The advantage of the targeted approach is that it is direct and therefore more rapid. However, it suffers from the limitation that it can only confirm or refute previous ideas and hence not reveal novel or unexpected features of pathogenicity. The disadvantage of the non-targeted approach is the scale of the experimentation required. It should be no surprise that the targeted approach has thus far yielded most results.

The targeted approach

The role of cutinase in overcoming the barrier to infection provided by cutin has been investigated by Kolattukudy and his colleagues (Kolattukudy, 1986; Woloshuk & Kolattukudy, 1986; Soliday *et al.*, 1988; Podila, Dickman & Kolattukudy, 1988; Dickman, Podila & Kolattukudy, 1989). Cutin is a waxy polymer of many fatty acids amongst which 10,16-dihydroxyhexadecanoic acid and 9,10,18-trihydroxyoctadecanoic acid are unique to it. The observation that some pathogens directly penetrate the cuticle suggests they must possess a cutinase. Amongst these was *Nectria haematococca* (= *Fusarium solani* f.sp. *pisi*) the cause of a pea root rot.

N. haematococca produces an abundant extracellular cutinase which is induced by the cutin monomers. This allowed the cDNA cloning of the cutinase gene and subsequent cloning of its genomic counterpart. Experiments with isolated nuclei indicated that the cutin monomers interact with a transacting factor to induce initiation of transcription of this gene. The most convincing experiment linking this gene to pathogenicity involved transforming it into *Mycosphaerella*, a pathogen of papaya which only infects wounded tissue. The transformant had the ability to infect intact papaya, with lesion size and lesion frequency being correlated with cutinase activity. Interestingly, isolated cutin monomers correctly induced expression in *Mycosphaerella* implying that this fungus possesses an analogue of the *N. haematococca* transacting factor.

Phytoalexins are antifungal compounds produced by plants in response to infection and successful pathogens must be able to tolerate the phytoalexin. An example of a well characterised system is the production of pisatin by peas in response to infection by *Nectria haematococca*. Van Etten, Matthews & Matthews (1989) have shown that strongly pathogenic strains are able to detoxify pisatin by demethylation to less toxic compounds. The gene for the P450 monooxygenase involved in demethylation (pisatin demethylase, PDA) was cloned by expression in *Aspergillus nidulans* (Weltring *et al.*, 1988). The critical role of this enzyme in *Nectria* pathogenicity was demonstrated by transformation of a strain which produced low amounts of PDA, with the cloned PDA gene resulting in the production of more virulent strains. More recently, Schafer *et al.* (1989) transformed the PDA gene into *Cochliobolus heterostrophus*, a leaf pathogen of maize. Transformants gained a pathogenic ability on pea leaves. Interestingly, the transformed strains were ineffective on pea roots whereas the native *N. haematococca* is ineffective on leaves. These results clearly demonstrate pathogenicity is dependent on PDA and that peas use pisatin as a general defence mechanism. The tissue specificity of transformed and wild-type strains points to an iceberg of complexity now within reach of experimental manipulation.

The non-targeted approach

An example of a non-targeted approach which has become increasingly focused is provided by the studies of De Wit on *C. fulvum*. *C. fulvum* is an intercellular pathogen and therefore molecules controlling pathogenicity and resistance must either be excreted into the intercellular fluid or be presented on cell surfaces. De Wit developed a technique to isolate intercellular fluid and has demonstrated that it contains plant and fungal proteins controlling the interaction (De Wit & Spikman, 1982). The first targets for investigation were the factors controlling avirulence and one of these, a polypeptide, has been purified and sequenced. (Scholtens-

Toma & De Wit, 1989). All available evidence suggests that it is the ultimate product of the avirulence gene 9. A fungal protein correlated with compatible interactions and plant proteins involved in the resistance response have also been purified (Joosten & De Wit, 1988a, b). It is clear that the analysis of intercellular fluids will provide much detailed information about avirulence, pathogenicity and resistance in the *C. fulvum*/tomato interaction.

The non-targeted approaches to pathogenicity also include the classical genetic approach of mutant isolation followed by complementation with gene libraries. Such an approach is attractive because of the wide range of genes which might be identified. Mutant isolation in most species is straightforward; a few species which are diploid or polyploid have been problematic. Mutants are selected on the basis of failure to cause disease despite normal vigour in axenic culture. Direct examination of the mutants can provide correlations with non-pathogenicity. Genetic analysis of plant pathogens is rarely straightforward because of the difficulty or even impossibility of sexual crossing. For this reason, we have developed a parasexual crossing system for *C. fulvum* involving protoplast fusion (Talbot *et al*., 1988). Transformation of mutants with gene libraries and selection of strains with restored pathogenicity is a formidable undertaking given that the assay for pathogenicity can be time and space consuming. It is also dependent on an efficient transformation system for cosmid libraries.

A second possible approach is to utilize promoter-probe vectors to select promoters switched on by growth in the plant. This approach has been applied successfully to bacterial pathogens (Osbourn, Barker & Daniels, 1987). The GUS system could form the basis of a promoter-probe vector and represents an ideal starting point for such a project. A third approach is to mutagenise strains and select for survivors whose pathogenicity is lost or altered. These non-pathogenic mutants should be prototrophic and of similar morphology to wild-type. Cloned gene banks can then be transformed into the mutants to identify gene(s) restoring pathogenicity to the mutant. One disadvantage of this approach is that it is rarely possible to demonstrate that only one gene is affected in the mutant. For this reason, the best mutagen to use may be the transforming vector itself. In this way, the disrupted gene is molecularly marked and can be identified by the inverse polymerase chain reaction (Ochman *et al*., 1990).

Conclusions

In this review we have attempted to demonstrate that the molecular analysis of plant fungal pathogenesis is beginning to bear fruit. In many areas such as transformation, gene expression and transposons, molecular studies of pathogens compare favourably with those of saprophytic fungi. Many enormous problems remain, perhaps the most important of which is how to tackle the experimental manipulation of the obligate pathogens, which include many of the economically most damaging fungi. Nevertheless, it is clear that the next few years will see a revolution in our understanding of the complexity and variety of pathogenic mechanisms leading perhaps to the development of novel methods of control.

References

Banks, G. R. (1983). Transformation of *Ustilago maydis* by a plasmid containing yeast 2-micron DNA. *Current Genetics* **7**, 73-77.

Banks, G. R. & Taylor, S. Y. (1988). Cloning of the PYR3 gene of *Ustilago maydis* and its use in DNA transformation. *Molecular Cell Biology* **8**, 5417-5424.

Bej, A. K. & Perlin, M. H. (1989). A high efficiency transformation system for the basidiomycete *Ustilago violacea* employing hygromycin resistance and lithium-acetate treatment. *Gene* **80**, 171-176.

Blakemore, E. J. A., Dobson, M. J., Hocart, M. J., Lucas, J. A. & Peberdy, J. F. (1989). Transformation of *Pseudocercosporella herpotrichoides* using 2 heterologous genes. *Current Genetics* **16**, 177-180.

Bond, T. E. T. (1938). Infection experiments with *Cladosporium fulvum* Cooke and related species. *Annals of Applied Biology* **25**, 277-307.

Churchill, A. C. L., Ciufetti, L. M., Hansen, D. R., Van Etten, H. D. & Van Alfen, N. K. (1990). Transformation of the fungal pathogen *Cryphonectria parasitica* with a variety of plasmids. *Current Genetics* **17**, 25-31.

Comino, A., Kolar, M., Schwab, H. & Socic, H. (1989). Heterologous transformation of *Claviceps purpurea*. *Biotechnology Letters* **11**, 389-392.

Cooley, R. N., Shaw, R. K., Franklin, F. C. H. & Caten, C. E. (1988). Transformation of the phytopathogenic fungus *Septoria nodorum* to hygromycin B resistance. *Current Genetics* **13**, 383-389.

Daboussi, M. -J., Djeballi, A., Gerlinger, C., Blaiseau, P. -L., Bouvier, I., Cassan, M., Lebrun, M. -H., Parisot, D. & Brygoo, Y. (1989). Transformation of seven species of filamentous fungi using the nitrate reductase gene of *Aspergillus nidulans*. *Current Genetics* **15**, 453-456.

De Wit, P. J. G. M. & Spikman, G. (1982). Evidence for the occurrence of race and cultivar-specific elicitors of necrosis in intercellular fluids of compatible interactions of *Cladosporium fulvum* and tomato. *Physiological Plant Pathology* **21**, 1-11.

De Wit, P. J. G. M., Toma, I. M. J. & Joosten, M. H. A.J. (1988). Race-specific elicitors and pathogenicity factors in the *Cladosporium fulvum* - tomato interaction. In *Physiology and Biochemistry of Plant Microbial Interactions*, ed. N. T. Keen, T. Kosuge & L. L. Walling, pp. 111-119. The American Society of Plant Physiologists: Rockville, USA.

Dickman, M. B. (1988). Whole cell transformation of the alfalfa pathogen *Colletotrichum trifolii*. *Current Genetics* **14**, 241-246.

Dickman, M. B., Podila, G. K. & Kolattukudy, P. E. (1989). Insertion of cutinase gene into a wound pathogen enables it to infect intact host. *Nature* **342**, 446-448.

Dinoor, A., Eshed, N. & Nof, E. (1988). *Puccinia coronata*, crown rust of oats and grasses. *Advances in Plant Pathology* **6**, 333-344.

Farman, M. L. & Oliver, R. P. (1988). The transformation of protoplasts of *Leptosphaeria maculans* to Hygromycin B resistance. *Current Genetics* **13**, 327-330.

Fotheringham, S. & Holloman, W. K. (1989). Cloning and disruption of *Ustilago maydis* genes. *Molecular and Cellular Biology* **9**, 4052-4055.

Glumoff, V., Kappeli, O., Fiechter, A. & Reiser, J. (1989). Genetic transformation of the filamentous yeast, *Trichosporon cutaneum*, using dominant selection markers. *Gene* **84**, 311-318.

Hansen, L. J., Chalker, D. L. & Sandmeyer, S. B. (1988). Ty3, a yeast retrotransposon associated with tRNA genes has homology to animal retroviruses. *Molecular and Cellular Biology* **8**, 5245-5256.

Henson, J. M., Blake, M. R. & Pilgeram, A. L. (1988). Transformation of *Gaeumannomyces graminis* to benomyl resistance. *Current Genetics* **14**, 113-117.

Jefferson, R. A., Kavanagh, T. A. & Bevan, M. W. (1987). GUS fusions: β-glucuronidase as a sensitive and versatile gene marker in higher plants. *EMBO Journal* **6**, 3901-3902.

Joosten, M. H. A. J. & De Wit, P. J. G. M. (1988a). Identification of several pathogenesis-related proteins in tomato leaves inoculated with *Cladosporium fulvum* (syn. *Fulvia fulva*) as 1,3-β-glucanases and chitinases. *Plant Physiology* **89**, 945-951.

Joosten, M. H. A. J. & De Wit, P. J. G. M. (1988b). Isolation, purification and preliminary characterisation of a protein specific for compatible *Cladosporium fulvum* (syn. *Fulvia fulva*)-tomato interactions. *Physiological and Molecular Plant Pathology* **33**, 142-153.

Kearney, B., Ronald, P. C., Dahlbeck, D. & Staskawicz (1988). Molecular basis for evasion of plant host defence in bacterial spot disease of pepper. *Nature* **322**, 541-543.

Kerr, E. A. & Bailey, D. L. (1964). Resistance to *Cladosporium fulvum* obtained from wild species of tomato. *Canadian Journal of Botany* **42**, 1541-1553.

Kolattukudy, P. E. (1986). Enzymatic penetration of the plant cuticle by fungal phytopathogens. *Annual Review of Phytopathology* **23**, 223-250.

Kronstad, J. W., Wang, J., Covert, S. F., Holden, D. W., McKnight, G. L. & Leong, S. A. (1989). Isolation of metabolic genes and demonstration of gene disruption in the phytopathogenic fungus *Ustilago maydis*. *Gene* **79**, 97-106.

Malardier, L., Daboussi, M. -J., Julien, J., Roussel, F., Scazzocchio, C & Brygoo, Y. (1989). Cloning of the nitrate reductase gene (*nia*D) of *Aspergillus nidulans* and its use for transformation of *Fusarium oxysporum*. *Gene* **79**, 147-156.

Mattern, I. E., Punt, P. J. & van den Hondel, C. A. M. J. J. (1988). A vector for *Aspergillus* transformation conferring phleomycin resistance. *Fungal Genetics Newsletter*, **35**, 25.

McHale, M. T., Roberts, I. N., Talbot, N. J. & Oliver, R. P. (1989). Expression of reverse transcriptase genes in *Fulvia fulva*. *Molecular Plant Microbe Interactions* **2**, 165-168.

Ochman, H., Medhora, M. M., Garza, D., Hartl, D. L. (1990). Amplification of flanking sequences by inverse PCR. In *PCR Protocols: A Guide to Methods and*

Applications, ed. M. A. Innis, D. H. Gelfrand, J. J. Sninsky & T. K. White, pp. 219-227. Academic Press: New York & London.

Oliver, R. P., Roberts, I. N., Harling, R., Kenyon, L., Punt, P. J., Dingemanse, M. A. & van den Hondel, C. A. M. J. J. (1987). Transformation of *Fulvia fulva*, a fungal pathogen of tomato, to hygromycin B resistance. *Current Genetics* **12**, 231-233.

Orbach, M. J., Vollrath, D., Davis, R. W. & Yanofsky, C. (1988). An electrophoretic karyotype for *Neurospora crassa*. *Molecular and Cellular Biology* **8**, 1469-1475.

Osbourn, A. E., Barber, C. E. & Daniels, M. J. (1987). Identification of plant-induced genes of the bacterial pathogen *Xanthomonas campestris* pathovar *campestris* using a promoter-probe plasmid. *EMBO Journal* **6**, 23-28.

Osiewacz, H. D. & Weber, A. (1989). DNA mediated transformation of the filamentous fungus *Curvularia lunata* using a dominant selectable marker. *Applied Microbiology and Biotechnology* **30**, 375-380.

Parsons, K. A., Chumley, F. G. & Valent, B. (1987). Genetic transformation of the fungal pathogen responsible for rice blast disease. *Proceedings of the National Academy of Sciences, USA* **84**, 4161-4165.

Podila, G. K., Dickman, M. B. & Kolattukudy, P. E. (1988). Transcriptional activation of a cutinase gene in isolated fungal nuclei by plant cutin monomers. *Science* **242**, 922-925.

Punt, P. J., Oliver, R. P., Dingemanse, M. A., Pouwels, P. H. & van den Hondel, C. A. M. J. J. (1987). Transformation of *Aspergillus* based on the hygromycin B resistance marker from *Escherichia coli*. *Gene* **56**, 117-124.

Roberts, I. N., Oliver, R. P., Punt, P. J. & van den Hondel, C. A. M. J. J. (1989). Expression of the *E. coli* β-glucuronidase gene in filamentous fungi. *Current Genetics* **15**, 177-180.

Saigo, K., Kugimiya, W., Matsuo, Y., Inouye, S., Yoshioka, K. & Yuk, S. (1984). Identification of the coding sequence for a reverse transcriptase-like enzyme in *Drosophila melanogaster*. *Nature* **312**, 659-661.

Schafer, W., Straney, D., Ciufetti, L., Van Etten, H. D. & Yoder, O. C. (1989). One enzyme makes a fungal pathogen, but not a saprophyte, virulent on a new host plant. *Science* **246**, 246-247.

Scholtens-Toma, I. M. J. & De Wit, P. J. G. M. (1989). Purification and primary structure of a necrosis-inducing peptide from apoplastic fluids of tomato infected with *Cladosporium fulvum* (syn. *Fulvia fulva*). *Physiological and Molecular Plant Pathology* **33**, 59-67.

Smyth, D. R., Kalitsis, P., Joseph, J. L. & Sentry, J. W. (1989). Plant retrotransposon from *Lilium henryi* is related to Ty3 of yeast and the Gypsy group of *Drosophila*. *Proceedings of the National Academy of Sciences, USA* **86**, 5015-5019.

Soliday, C. L., Flurkey, W. H., Okita, T. W. & Kolattukudy, P. E. (1988). Cloning and structure determination of cDNA for cutinase, an enzyme involved in fungal penetration of plants. *Proceedings of the National Academy of Sciences, USA* **81**, 3939-3943.

Talbot, N. J., Coddington, A., Roberts, I. N. & Oliver, R. P. (1988). Diploid construction by protoplast fusion in *Fulvia fulva* (syn. *Cladosporium fulvum*): genetical analysis of an imperfect plant pathogen. *Current Genetics* **14**, 567-572.

Tsukuda, T., Carlton, S., Fotheringham, S. & Holloman, W. K. (1988). Isolation and characterisation of an autonomously replicating sequence from *Ustilago maydis*. *Molecular and Cellular Biology* **8**, 3703-3709.

Turgeon, B. G., Garber, R. L. & Yoder, O. C. (1985). Transformation of the fungal maize pathogen *Cochliobolus heterostrophus* using the *Aspergillus nidulans amdS* gene. *Molecular and General Genetics* **201**, 450-453.

Turgeon, B. G., Garber, R. L. & Yoder, O. C. (1987). Development of a fungal transformation system based on selection of sequences with promoter activity. *Molecular and Cellular Biology* **7**, 3297-3305.

Van Etten, H. D., Matthews, D. E. & Matthews, P. S. (1989). Phytoalexin detoxification: importance for pathogenicity and practical implications. *Annual Review of Phytopathology* **27**, 143-164.

Weltring, K. M., Turgeon, B. G., Yoder, O. C. & Van Etten, H. D. (1988). Isolation of a phytoalexin detoxification gene from the plant pathogenic fungus *Nectria haematococca* by detecting its expression in *Aspergillus nidulans*. *Gene* **68**, 335-344.

Woloshuk, C. P. & Kolattukudy, P. E. (1986). Mechanism by which contact with plant cuticle by fungal phytopathogens triggers cutinase gene expression in the spores of *Fusarium solani* f.sp. *pisi*. *Proceedings of the National Academy of Sciences, USA* **83**, 1704-1708.

Index

A

Absidia glauca 9
acetamidase 106
acetyl CoA synthetase gene 36
Achlya ambisexualis 108
acid phosphatase gene 68
Acremonium chrysogenum 35
alcohol dehydrogenase 103
alcohol dehydrogenase gene 67, 69
alcohol oxidase 119
allyl alcohol 110
am gene 35
ampicillin resistance 32
amylase 75, 107, 166
amyloglucosidase 138
amylolytic yeast 167
antibiotic biosynthesis genes 36
antithrombin III 75
antitrypsin 75
areA gene 35
Ascobolus immersus 9 - 10
aspartyl proteinase 107
Aspergillus awamori 107
Aspergillus expression signals 15
Aspergillus nidulans 1, 7, 9, 15 - 16, 29 - 30, 32 - 36, 103, 150
Aspergillus niger 12 - 16, 18, 36, 103, 107, 138
Aspergillus oryzae 107
ATP synthetase gene 35
autonomous replication sequences 72
autonomously replicating vectors 9
auxotrophic mutants, in transformant selection 3
auxotrophic selectable markers 4 - 5
avirulence genes 170

B

Bacillus subtilis 131
beer 131
benomyl resistance 4 - 7
benzoate hydroxylase 11
benzoate hydroxylase, gene cloning 12, 14
biolistic 44
brewing 131

C

calf chymosin 94
cartridge gun 46
catabolite repression 68
catalase 165
cellobiohydrolase 75, 85, 86
cellobiohydrolase gene 107
cellulase 85
cellulase genes 87
cellulase genes, inactivation 93
cellulase mutants 91
cellulase, induction 86
cellulase, structure & function 87 - 90
cellulases, industrial uses 86
centromere regions 72
cephalosporin 1, 146
Cephalosporium acremonium 108, 150
cephamycin 146
chlorate resistance 3, 171
chromosome aberrations 34
chromosome walking 36 - 37
chymosin 94, 95 - 97, 107, 111
chymosin production 18, 21
citric acid 1
Cladosporium fulvum 170

cloned genes, identification 37
cloning strategies 29 - 31, 33, 35, 37, 39, 41, 43
Cochliobolus heterostrophus 174
Colletotrichum trifolii 172
colony stimulating factor 106
commercial production 108
commercial products 1
complementation, in *E. coli* 31
complementation, in filamentous fungi 32
complementation, in yeast 32
copper resistance 134
Coprinus cinereus 9, 32, 37
cosmid clone 12
cosmid libraries 32
cosmid vectors 30 - 31, 37
crop losses, due to pathogenic fungi 1
cytochrome *c* gene 69

D

degradation, by proteinase 113
differential hybridization 34
disintegration vector 72, 137
dominant selectable markers 3, 8, 71, 134 - 135

E

ectopic integration 9, 14, 38
electric gun 45, 48 - 49
electroporation 3, 44, 51, 56
electroporation chamber 54
electroporation, apparatus design 53
endoglucanase 85, 105
endoglucanase enrichment 91
endoglucanase gene 86, 131
erythropoietin 75
ethanol regulon 110
exoglucanases 85
expression cassette 66, 103, 168
expression cassette, stability 71
expression signals 15
expression system 161-162
expression vector 15, 69, 108, 121, 162
expression vectors, *Trichoderma* 94
expression, of galactosidase 124
expression, heterologous genes 104

F

filtration, glucanase enhancement 131
flavour 131
fluoro-orotic acid resistance 3
formate dehydrogenase gene 162
Fusarium solani 176

G

galactose metabolism genes 67
galactosidase 121, 124
galactosidase genes 15
galactosidase promoters 15
gene disruption 13, 38, 156, 172, 178
gene fusion 18, 21, 108, 167
gene integration 138 - 139
gene isolation 31
gene replacement 3, 9, 92
gene transfer 2
glucoamylase 1, 75, 103, 162, 166 - 167
glucosidases 85
glucuronidase gene 174
glutamate dehydrogenase 35
glyceraldehyde phosphate dehydrogenase gene 67
glycosylation 75 - 76, 125 - 126, 161
guar plant 121
gun, cartridge 46
gun, electric 45, 48 - 49
gun, particle 44 - 45

H

Hansenula polymorpha 119, 161
hepatitis surface antigen 77, 163
heterologous expression 29, 31, 66, 104, 118
heterologous probes 35
heterologous protein expression 126
heterologous protein production 18, 103

heterologous protein,
produced by *Trichoderma* 93
heterologous proteins 163
histone genes 35
homologous integration 15
homologous recombination 9, 172
homologous targeting 104
horizontal transfer 36, 149, 151, 154
human interferon 105, 108
human serum albumin 69, 75
hybrid promoters 69 - 70
hybrid release translation 34
hybrid VLPs 78
hygromycin B resistance 7
hyperglycosylation 75, 94, 112 - 113

I

identification of cloned genes 37
in vitro mutagenesis 140
inactivation, of cellulase genes 93
increased production, of
endoglucanase 91
integration 104, 122
integration events 9
integration vector 162
integrative transformation 137,
68, 172
intercellular fluids 178
intergenic promoter regions 16
intracellular production 76
isocitrate lyase gene 36

K

karyotype, molecular 133, 174

L

lactam antibiotics 146
lactam biosynthesis 149
leader sequence 74
Leptosphaeria maculans 172
Lycopersicon 170
lysozyme 75

M

marker rescue 33
markers, selectable 3, 71, 173
mating type factors 37
maturation, of proteins 75
methanol oxidase 164
methanol oxidase gene 162, 165
methionylaminopeptidase 77
methylotrophic yeasts 118, 161
microprojectiles 44, 45
mitochondrial genomes 29
molecular karyotyping 133, 174
Mucor circinelloides 9
Mucor miehei 107 - 108
mutagenesis, *in vitro* 140

N

Nectria haematococca 9, 32, 176
Neurospora crassa 1, 9, 29 - 30,
32 - 33, 35 - 36, 108
non-homologous recombination 9

O

OFAGE 133
oligonucleotide probes 35

P

particle gun 44 - 45
pathogenicity genes 176
penicillin 1, 146
Penicillium chrysogenum 35 - 36,
38, 108, 150
peroxisomes 161
phleomycin resistance 7, 14, 173
phosphoglycerate kinase gene 35, 67
Phycomyces blakesleeanus 9
phytoalexins 177
Pichia polymorpha 119
pisatin demethylase 32
plasma membrane, electrical
breakdown 52
plasmid rescue 32
plasmid stability 71, 135 - 136

plasmid vectors 31
plasmids, of *Nectria* 9
plasminogen activator 106
ploidy of yeast 132, 133
Podospora anserina 9, 32
polyethylene glycol 44
polymorphism, yeast rDNA 133
positive selection 3
processing, of leader sequence 74
prochymosin 18, 21, 68, 105 - 106
prochymosin, production 18, 21
production of heterologous
 proteins 103
production, commercial 108
products, commercial 1
promoter-probe vectors 178
promoters 15
promoters, from yeast 66 - 67
promoters, hybrid 69 - 70
protease gene 69
protein maturation 75
protein processing 74
protein secretion 73
proteinase degradation 113
protoplast formation 2
protoplast fusion 178
protoplasts 44, 171
pulsed field electrophoresis 30, 174
pyr-4 gene 36
pyrG gene 36
pyrogenicity 66

Q

qa genes 36
Quorn 1

R

rDNA polymorphism 133
recipient strain, improvement 121
regulon, ethanol 110
reporter genes 15, 18, 173
resistance, to copper 134
retrotransposon 175
RFLP 134

S

Saccharomyces cerevisiae 9, 31, 35, 118, 132
Saccharomyces monacensis 134
Schwanniomyces occidentalis 166
secretion efficiency 76
secretion of proteins 73
secretion pathway 112 - 113
secretion signals 103
secretion, of galactosidase 124
selectable markers 3, 71, 173
 71, 134 - 135
sib-selection 33
signal peptide, cleavage 112
signal sequences, synthetic 105
single chromosome transfer 133
soy sauce 1
stability of expression cassettes 71
stability, of plasmids 135, 136
stability, of transformants 104 - 105, 123, 162
strain improvement, in yeast 129, 130
suicide enrichment 122
suppressor tRNA gene as selectable
 marker 3
synthetic signal sequences 105

T

terminators 70
Tetrahymena thermophila 9
tomato 170
transcription terminators 70
transformant, stability 104-105, 162
transformants, multiple-copy 104
transformation 32, 134, 171
transformation procedures 2
transformation systems 2
transforming DNA, fate 9
transposon 175
Trichoderma reesei 85, 107, 131
Trichoderma, expression vectors 94
Trichoderma, producing
 heterologous protein 93
Trichosporon cutaneum 9
triosephosphate isomerase 105

tubulin 4 - 7
tubulin genes 35
tumour necrosis factor 119
Ty element 175
Ty element, gene fusions 78

U

UAS, upstream activation sequence 67
ubiquitin gene 77
urokinase plasminogen activator 75
Ustilago maydis 9, 36
Ustilago violacea 172

V

vector development 1
vectors 9, 31
virulence 170
virus like particles 78

X

xylanase 85

Y

yeast transformation 134
yeast, ploidy 132 - 133
yeast, strain improvement 129 - 130